MW01285781

Dinah Zike's

Big Book of

Graphics

Every Teacher Needs!

By Dinah Zike, M.Ed.

dma
dinah-might activities

© 1999 Dinah Zike

Book design and layout: D.ig Productions, Comfort, Texas
Production: Dinah Zike, Ignacio Salas-Humara
Photography: © 1999 Ignacio Salas-Humara
Illustrations: Chris Holden, Jesse Flores, Noe Garza, and Ignacio Salas-Humara
© 1999 Dinah-Might Activities, Inc.
Editor: Ignacio Salas-Humara
Printed by Signature Graphics, San Antonio, Texas

Although the contents of this book are copyrighted, certain portions
are meant to be photocopied for use by the purchaser in the classroom and/or home.

Other books and videos by Dinah Zike:

The Big Book of Books and Activities
The Big Book of Projects
The Big Book of Holiday Activities
Great Tables, Graphs, Charts, Diagrams, and Timelines You Can Make!
Science Poems, Riddles, and Rhymes: Earth, Solar System, and Universe
Science Poems, Riddles, and Rhymes: Land, Water, Air, and Weather
The Earth Science Book
Video: How to Use the Big Book of Books
Video: How to Use the Big Book of Projects
Time Twister Chronicles: The Search For T. Rex
Time Twister Chronicles: Rainforest Rescue
Time Twister Chronicles: The Hidden Caverns

Dinah's books are available in teacher bookstores or can be ordered from:

Dinah-Might Activities, Inc.
P.O. Box 690328
San Antonio, Texas, 78269
Office: (210) 698-0123
Fax: (210) 698-0095
Orders: 1-800-99-DINAH (993-4624)

Table of Contents

A Note From Dinah Zike

The math graphics featured in this book are compiled from the following sources: graphics from my discontinued "booklet" series; graphics I used in my classrooms when I was teaching; and new graphics I have developed while working with my adopted schools. These duplicable pages are very generic, and can be used with any math program. Those of you who are familiar with my *Big Book of Books and Activities* will note that some of the pages are presented in a particular arrangement with a specific number of graphics on a page so they can be used with my folds and graphic organizers. When this is the case, I have named the folded book that the graphics accompany.

I am currently working on a *Big Book of MATH ACTIVITIES* that will combine all of my math ideas, hints, activities, graphic organizers, and student-made books into one big math book. This book is scheduled be published Fall 2000. If you would like to be on a mailing list for this book, please call 1-800-99DINAH or fax your mailing information to (210) 697-0095. We will also send you a catalog of other Dinah-Might Activities publications upon request.

I hope to see you in workshops or at conferences soon!

For the Love of Learning,

Dinah Zike

10/99

Money Graphics

Large Half Dollar: Front

Large Half Dollar: Back

SCHOOL MONEY

INDEPENDENCE

HALF DOLLAR

Large Quarter: Front

Large Quarter: Back

Large Dime: Front and Back

Large Nickel: Front and Back

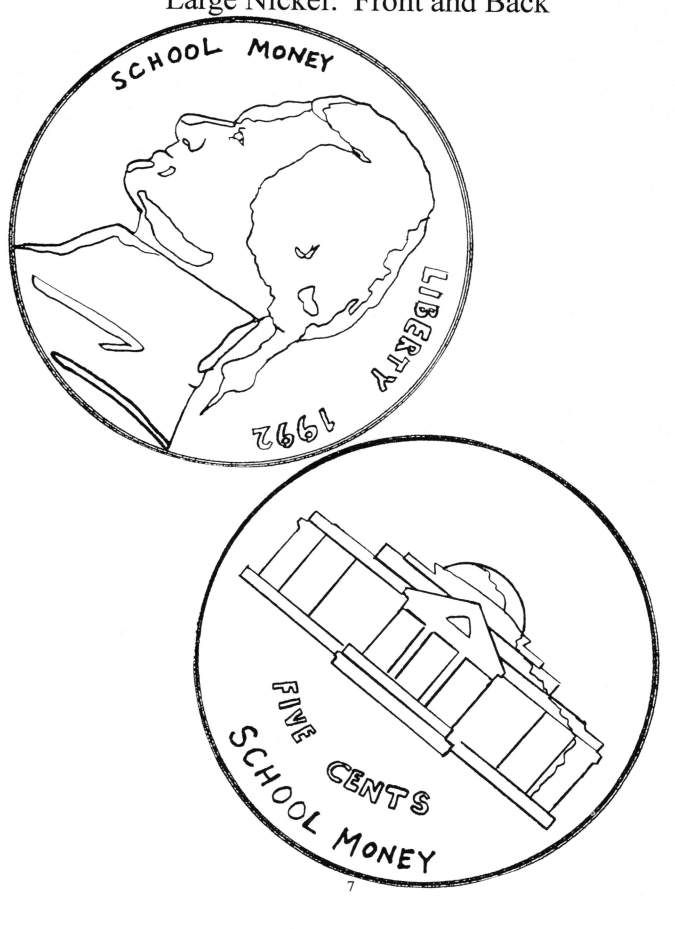

Large Penny: Front and Back

Small Money Pieces: Mixed

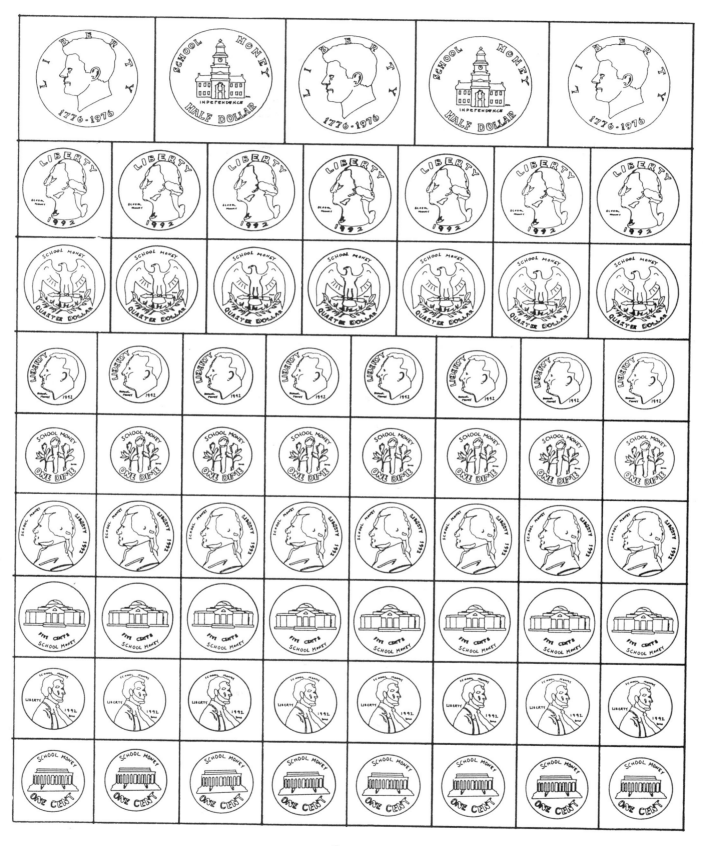

Small Half Dollars

Small Quarters

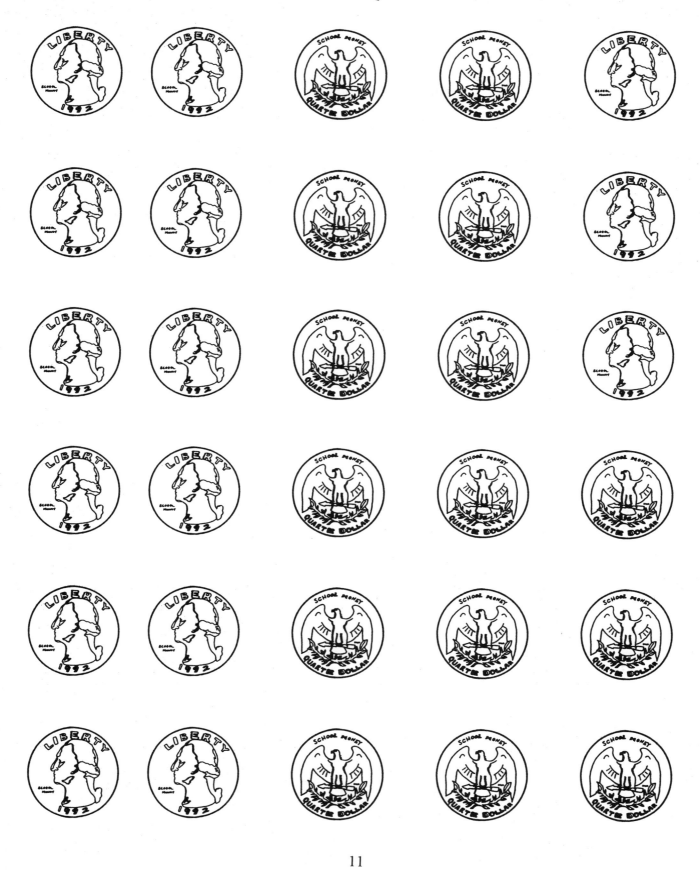

Small Dimes

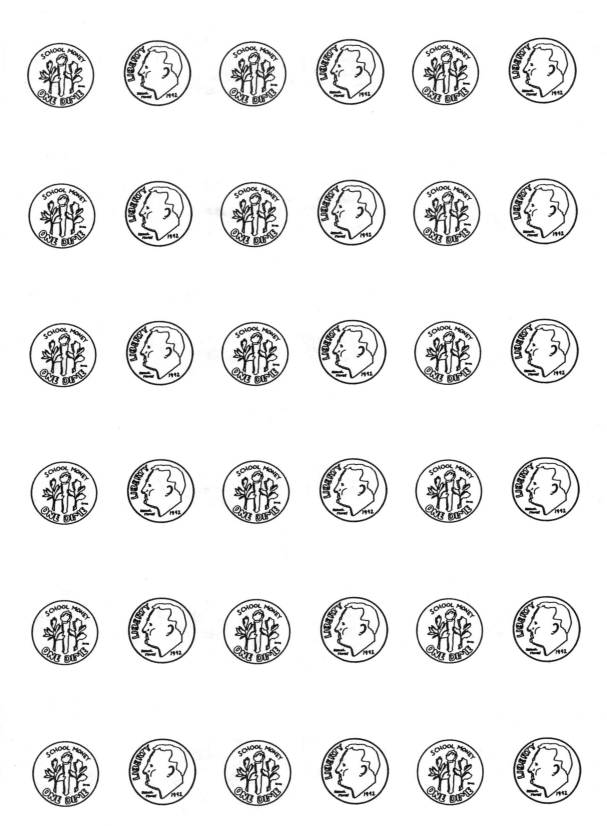

Small Nickels

Small Pennies

Twenty Dollar Bill: Front and Back

Ten Dollar Bill: Front and Back

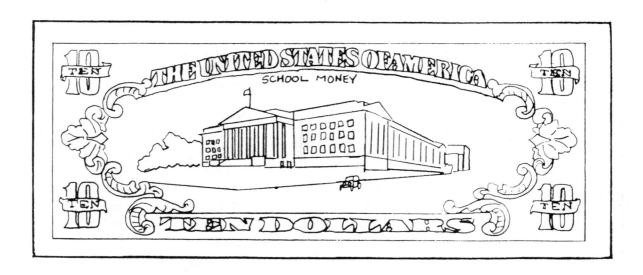

Five Dollar Bill: Front and Back

One Dollar Bill: Front and Back

Money Combinations

Money Match Book

Glue different combinations of coins inside this money chest and calculate how much money is inside.

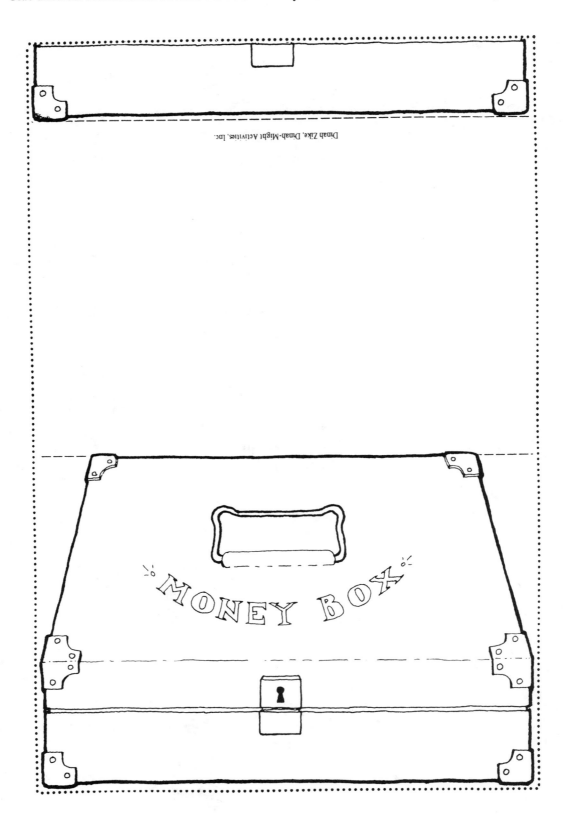

Dinah Zike, Dinah-Might Activities, Inc.

Money Equivalencies:
Use With Layred Look Book

Time

Large Clock Face

Large Clock Face:
Roman Numerals

Hour, Minute, and Second

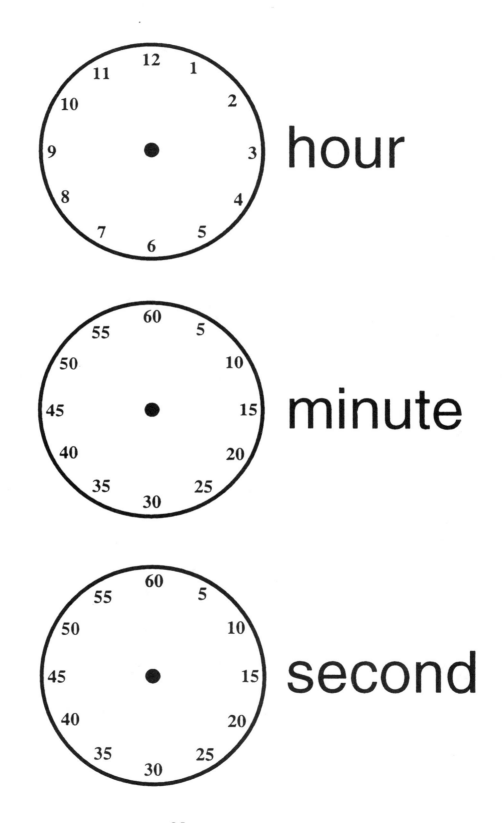

Small Clock Faces

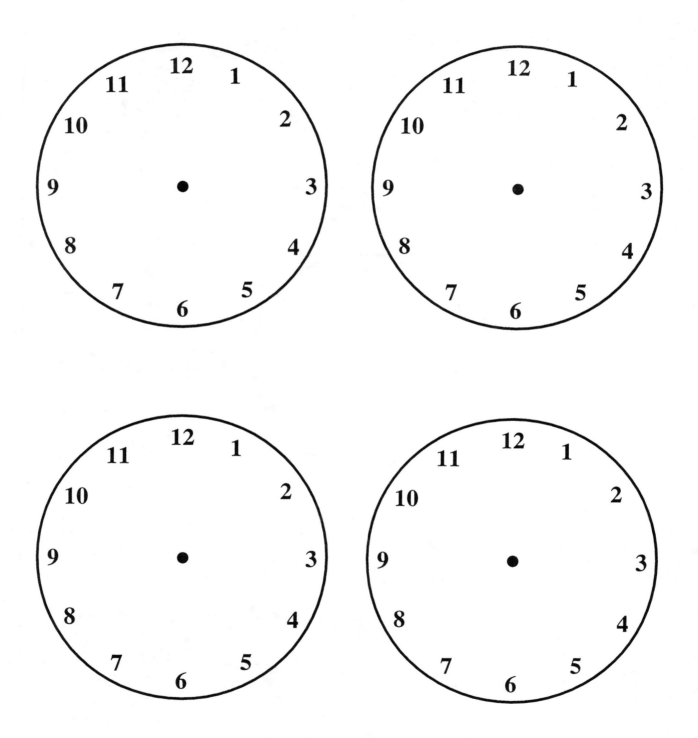

Small Clock Faces

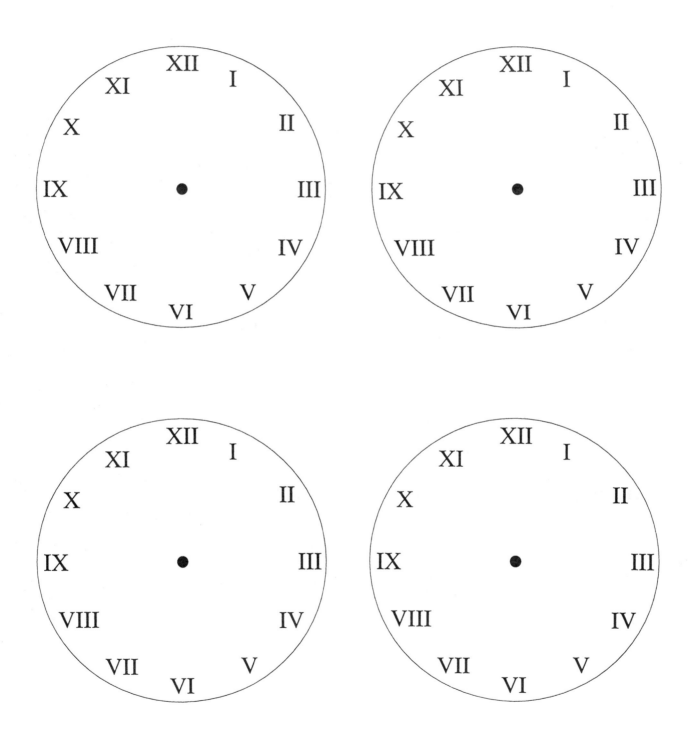

Small Clock Faces: Military Time

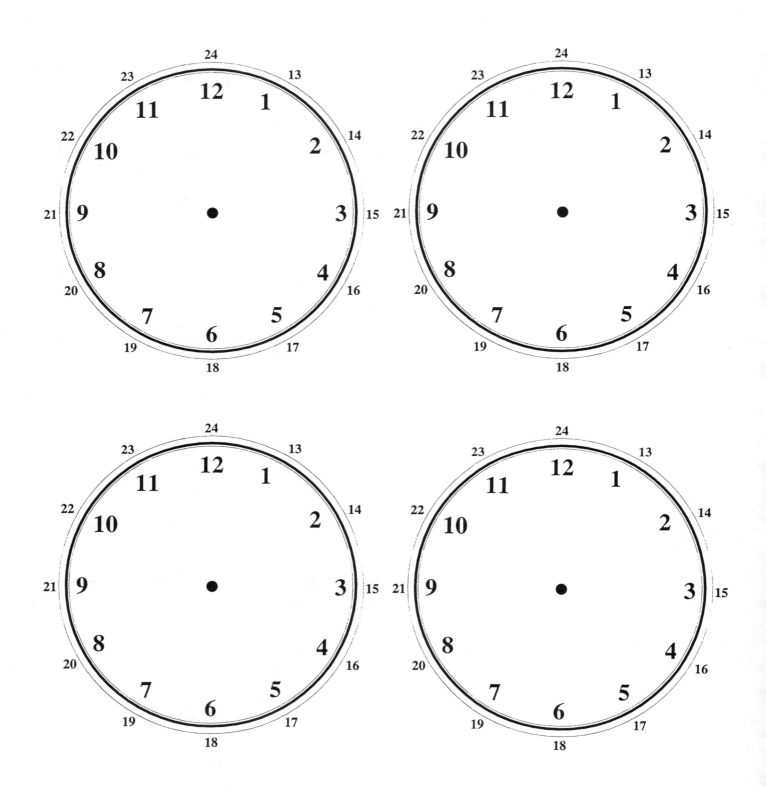

Small Clock Faces:
Use With Layered Look Book

Clock Face:
Use Inside Envelope Fold

12:00	5:00
12:30	5:30
1:00	6:00
1:30	6:30
2:00	7:00
2:30	7:30
3:00	8:00
3:30	8:30
4:00	9:00
4:30	9:30

10:00	5:11
10:30	5:55
11:00	6:33
11:30	7:29
12:08	8:16
12:47	9:01
1:15	10:52
2:43	11:20
3:05	12:12
4:39	1:10

Century, Decade, Years:
Use With Layered Look Book

CENTURY

DECADE | DECADE | DECADE | DECADE | DECADE | DECADE | DECADE | DECADE | DECADE

10 YEARS | 10 YEARS | 10 YEARS | 10 YEARS | 10 YEARS | 10 YEARS | 10 YEARS | 10 YEARS | 10 YEARS

Year

Month	Month	Month	Month	Month	Month	Month	Month	Month	Month	Month	Month
1	2	3	4	5	6	7	8	9	10	11	12
January	February	March	April	May	June	July	August	September	October	November	December

Jan.	Feb.	Mar.	April	May	June	July	Aug.	Sept.	Oct.	Nov.	Dec.
31 Days	28 or 29 Days	31 Days	30 Days	31 Days	30 Days	31 Days	31 Days	30 Days	31 Days	30 Days	31 Days

Large Calendar

Month:						
S	**M**	**T**	**W**	**T**	**F**	**S**

Small Calendars: Publishing Center

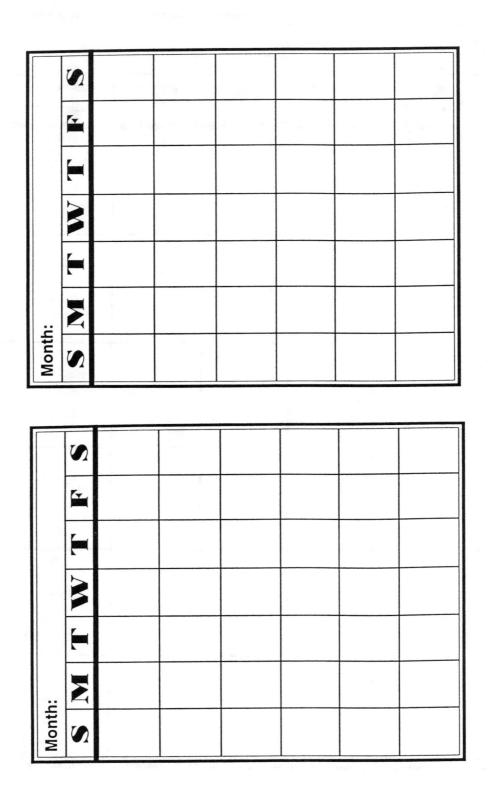

Millennium, Century, Decade:
Use Inside Envelope Fold

MILLENNIUM

| CENTURY | CENTURY | CENTURY | CENTURY | CENTURY | CENTURY | CENTURY | CENTURY | CENTURY |

| 10 DECADES | 10 DECADES | 10 DECADES | 10 DECADES | 10 DECADES | 10 DECADES | 10 DECADES | 10 DECADES | 10 DECADES |

Large Match Book Writing Activity

Draw hands on the clock to illustrate a time. Write inside the book about what happened at that time.

Dinah Zike, Dinah-Might Activities, Inc.

Trifold Clock Writing Activity

Draw hands on the clock to illustrate a time. Write inside the book about what happened at that time.

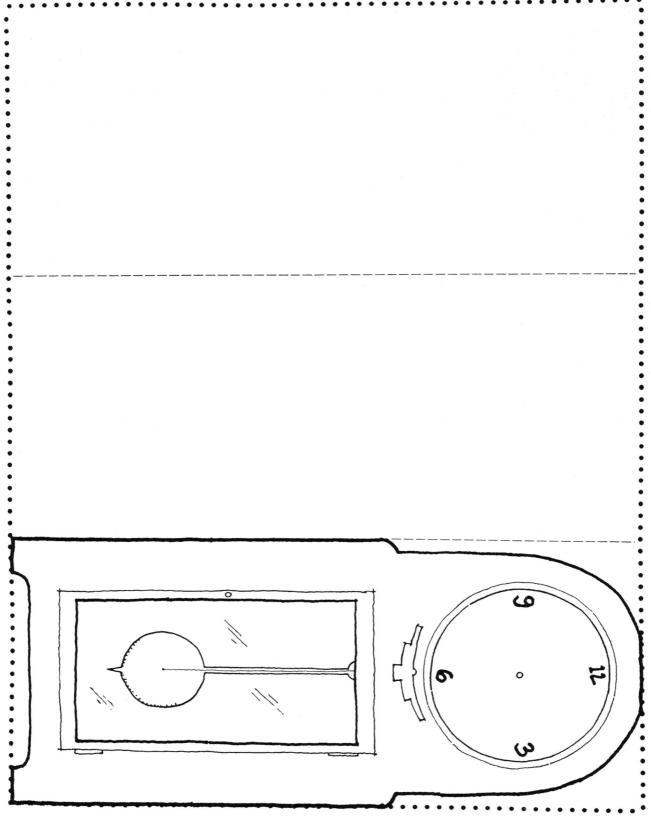

Measurement

6 Inch Rulers
Use With Layered Look Book

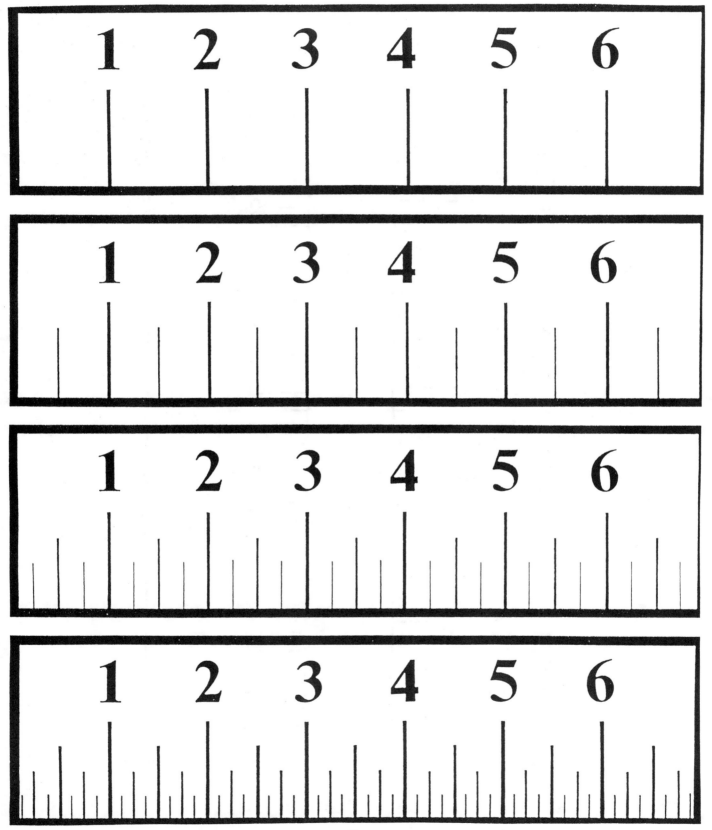

Metric Ruler

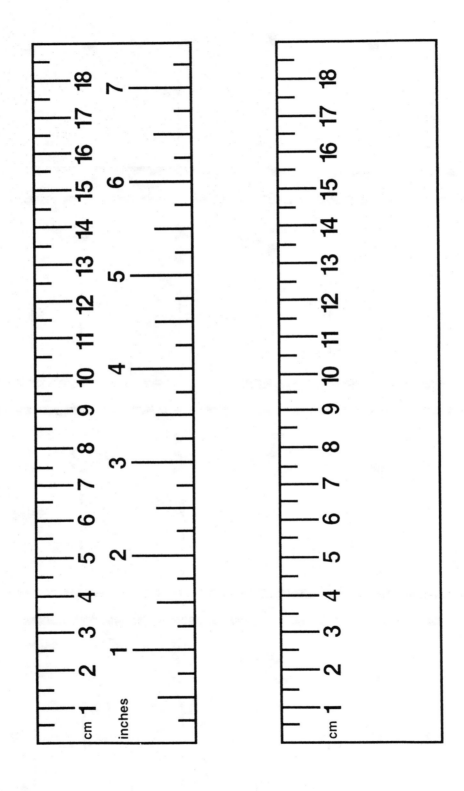

Miles and Feet:
Use With Layered Look Book

MILE

1/2 MILE

1/2 MILE

1/4 MILE

1/4 MILE

1/4 MILE

1/4 MILE

1/8 MILE

1/8 MILE

1/8 MILE

1/8 MILE

1/8 MILE

1/8 MILE

1/8 MILE

1/8 MILE

Pounds and Ounces:
Use With Layered Look Book

ONE POUND
16 OUNCES

1/2 POUND	1/2 POUND
8 OUNCES	8 OUNCES

1/4 POUND	1/4 POUND	1/4 POUND	1/4 POUND
4 OUNCES	4 OUNCES	4 OUNCES	4 OUNCES

1/8 LB	1/8 LB	1/8 LB	1/8 LB	1/8 LB	1/8 LB	1/8 LB	1/8 LB
2 OUNCES	2 OUNCES	2 OUNCES	2 OUNCES	2 OUNCES	2 OUNCES	2 OUNCES	2 OUNCES

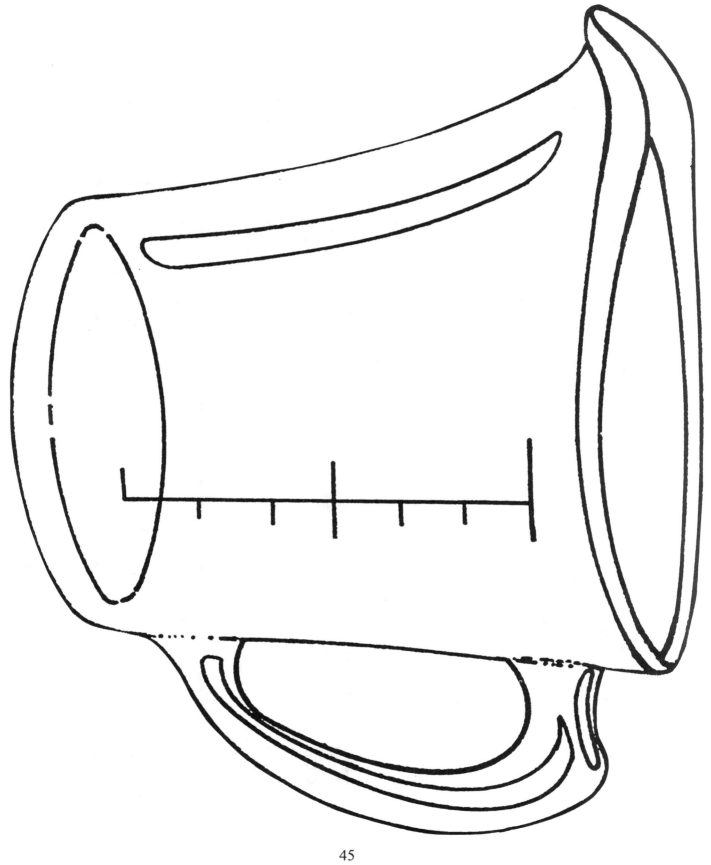

Cups and Ounces:
Use With Layered Look Book

ONE CUP

8 OUNCES

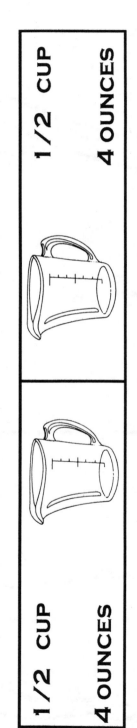

1/2 CUP

4 OUNCES

1/2 CUP

4 OUNCES

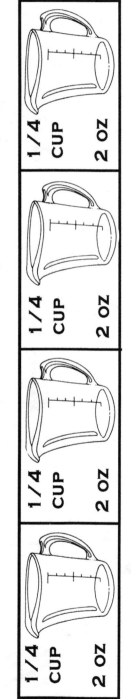

1/4 CUP

2 OZ

1/4 CUP

2 OZ

1/4 CUP

2 OZ

1/4 CUP

2 OZ

1/8 CUP 1OZ	1/8 CUP 1OZ	1/8 CUP 1OZ	1/8 CUP 1OZ	1/8 CUP 1OZ	1/8 CUP 1OZ	1/8 CUP 1OZ	1/8 CUP 1OZ

Cups In a Gallon

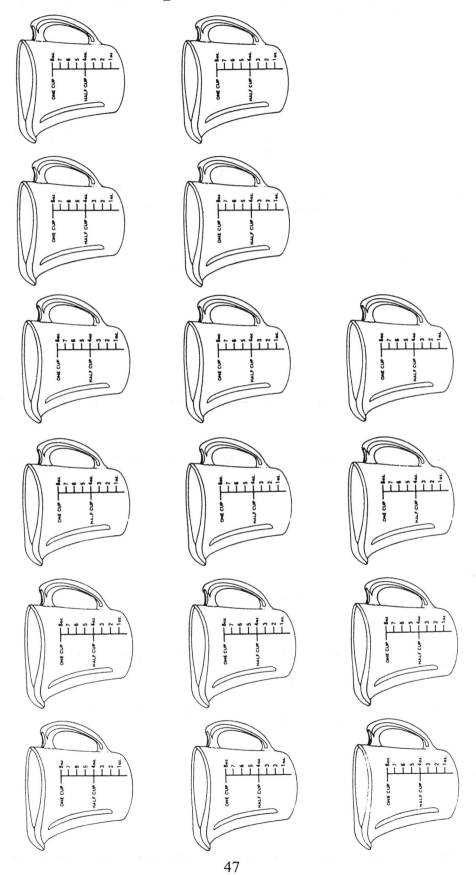

Classroom Beaker: Bulletin Board

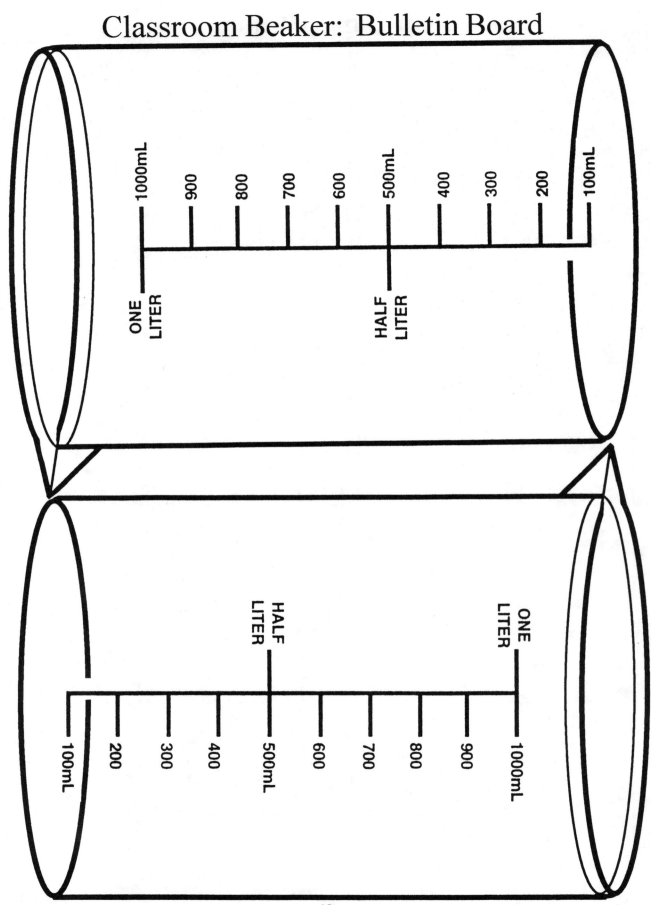

Classroom Flask: Bulletin Board

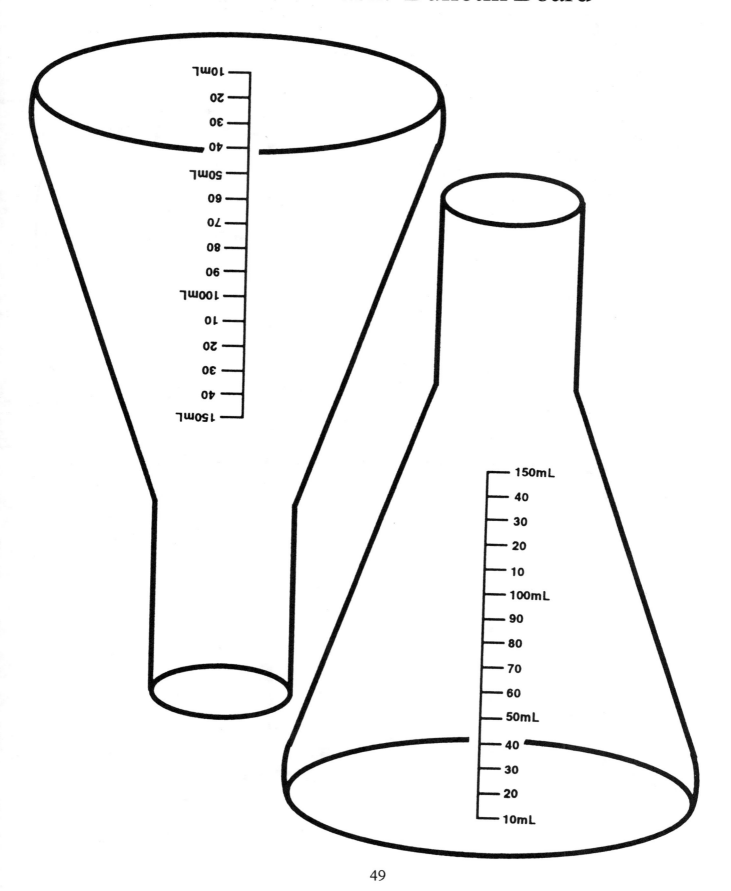

Classroom Cylinder: Bulletin Board

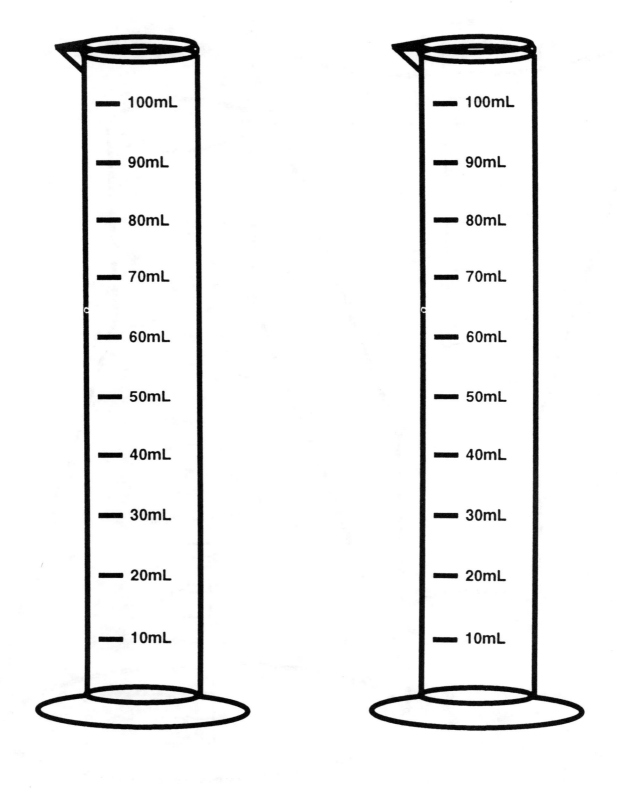

Student Beakers: Publishing Center

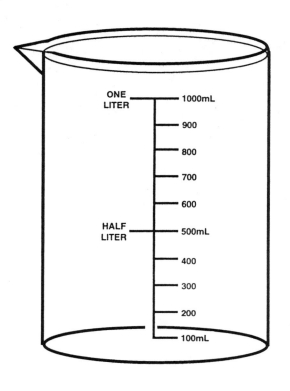

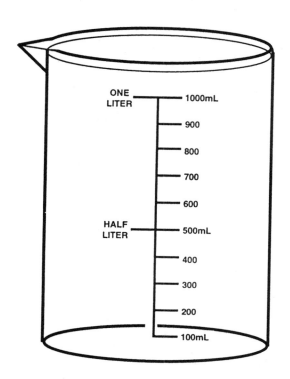

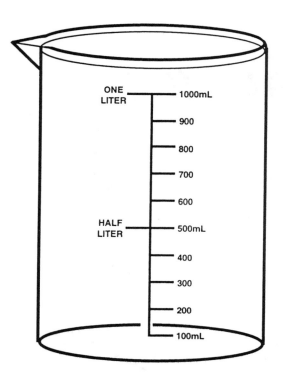

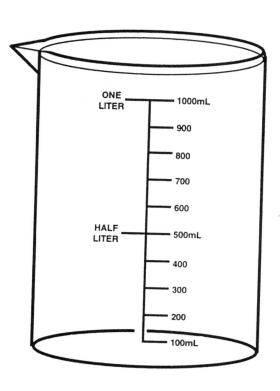

Classroom Gallon: Bulletin Board

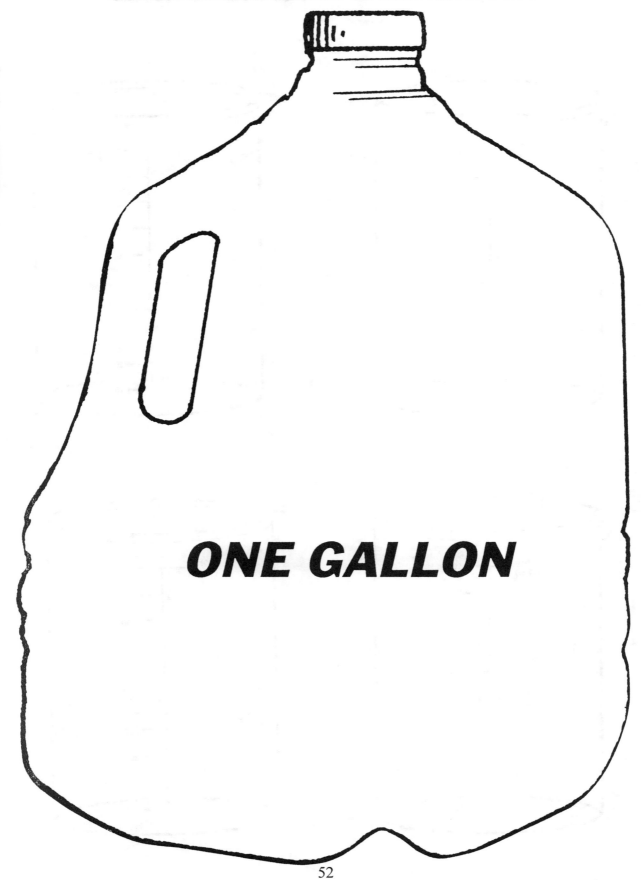

ONE GALLON

Classroom Quarts: Bulletin Board

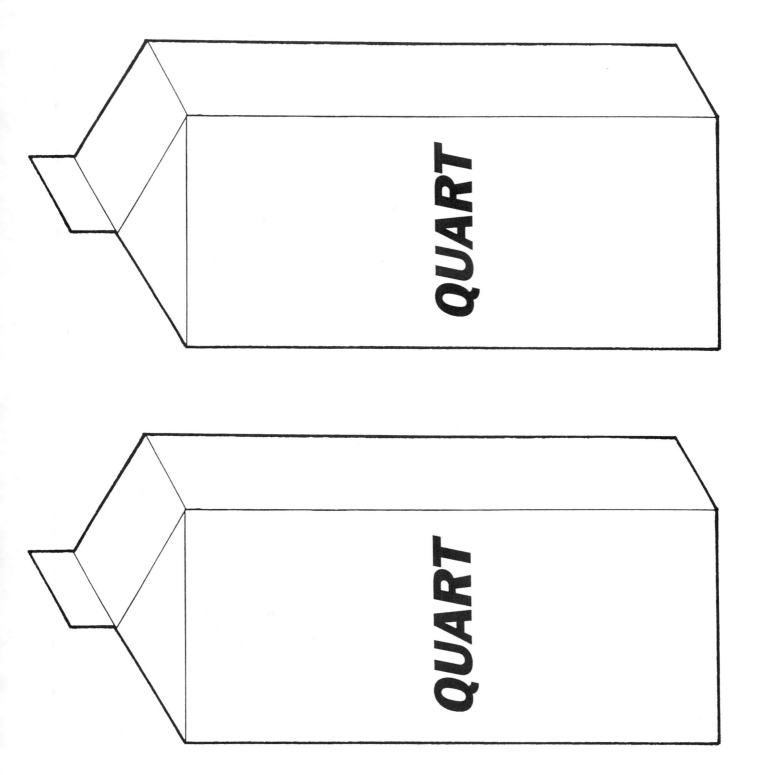

Classroom Pints: Bulletin Board

Student Patterns

Liquid Measurement
Use With Layered Look Book

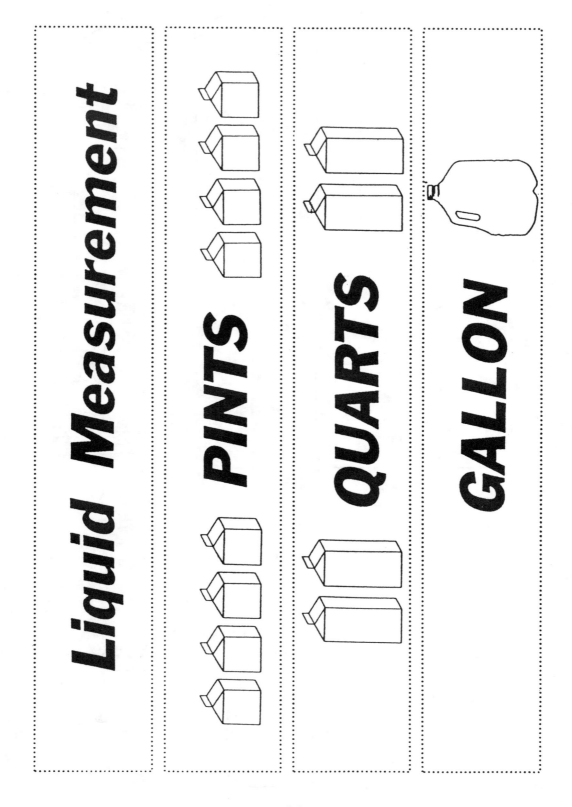

Liquid Measurement

PINTS

QUARTS

GALLON

Weather Measurement

Student Raingauges: Publishing Center

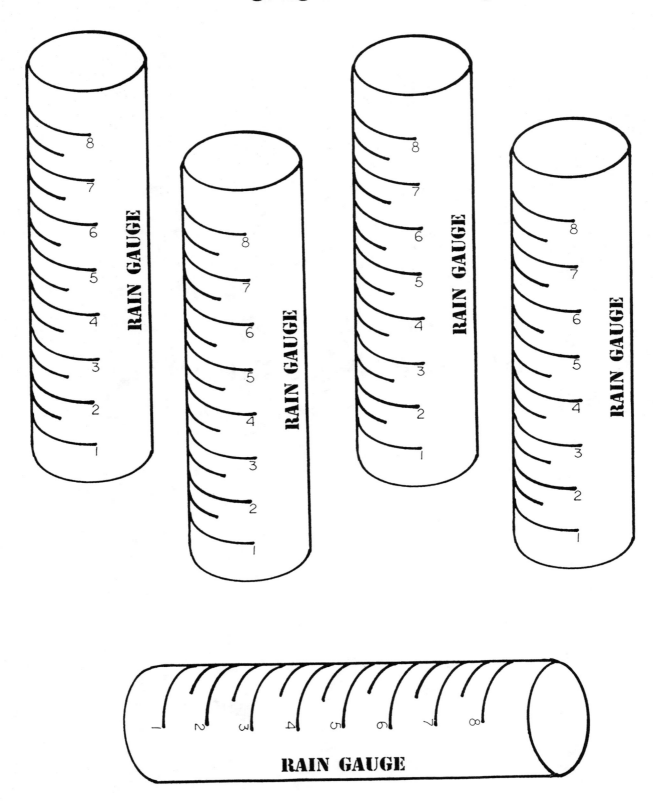

Classroom Raingauges: Bulletin Board

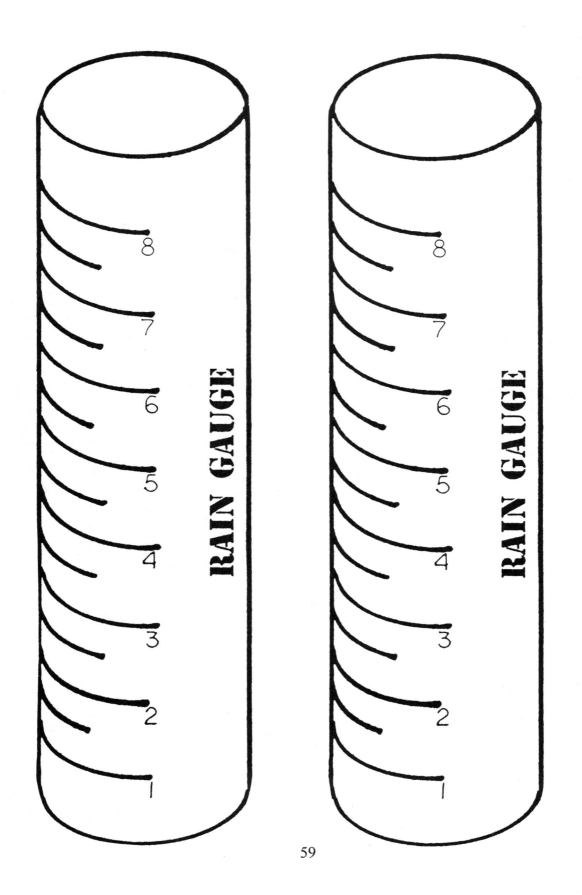

Classroom Thermometers

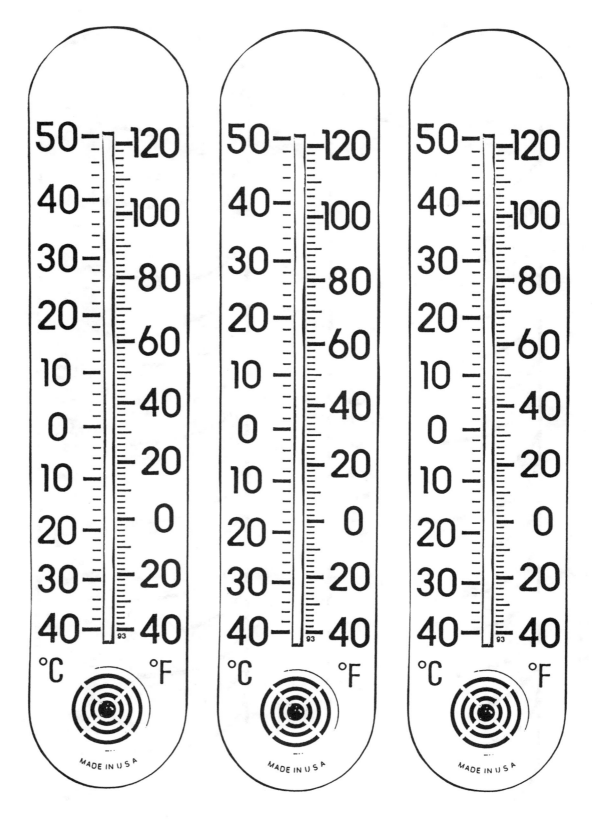

Student Thermometers: Publishing Center

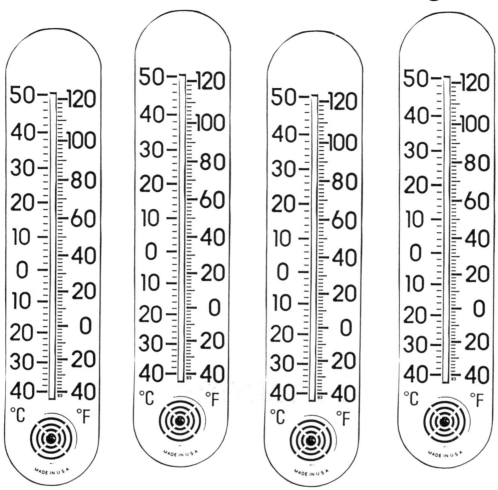

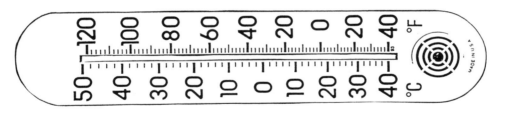

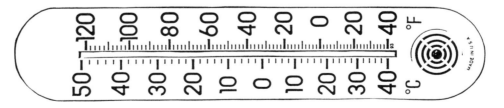

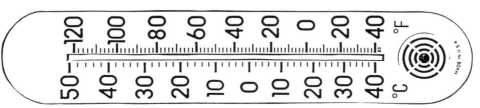

Fractions

Circles: Fractional Parts

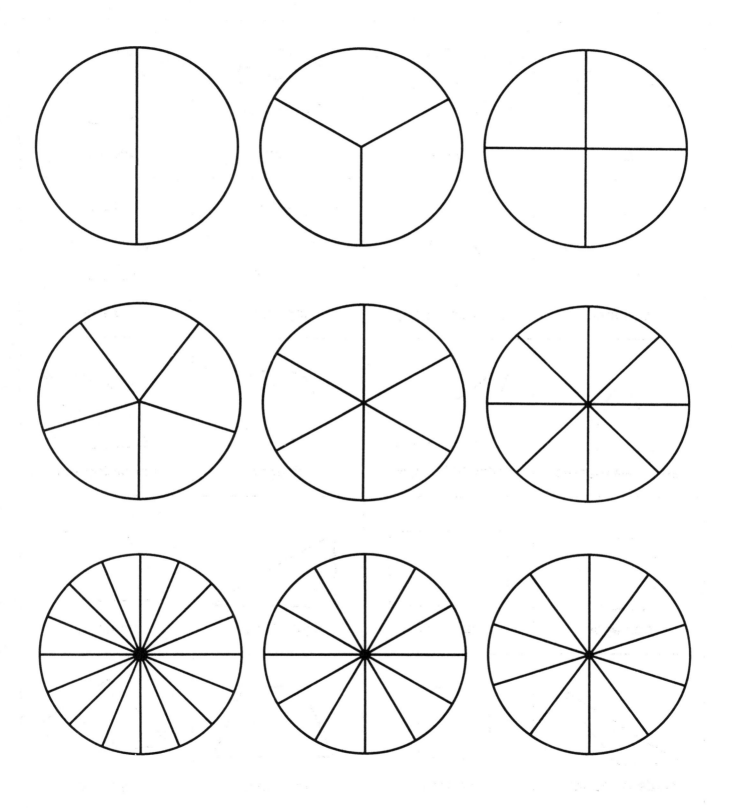

Shapes: To Be Divided Into Fractional Parts

arch	circle	cube	decagon
diamond	equilateral triangle	heptagon	hexagon
isosceles triangle	obtuse triangle	octagon	parallelogram
pentagon	rectangle	right triangle	rhombus
scalene triangle	square	trapezoid	pyramid

Squares: Fractional Parts

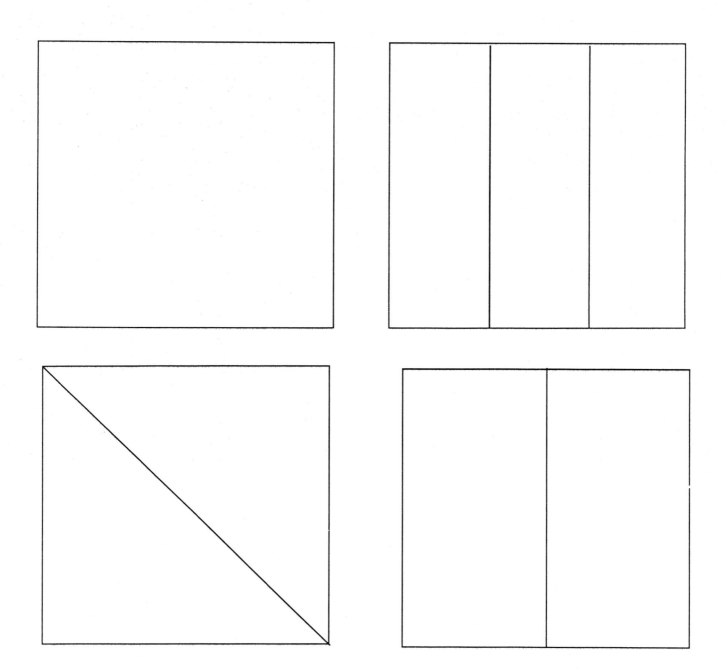

Squares: Fractional Parts

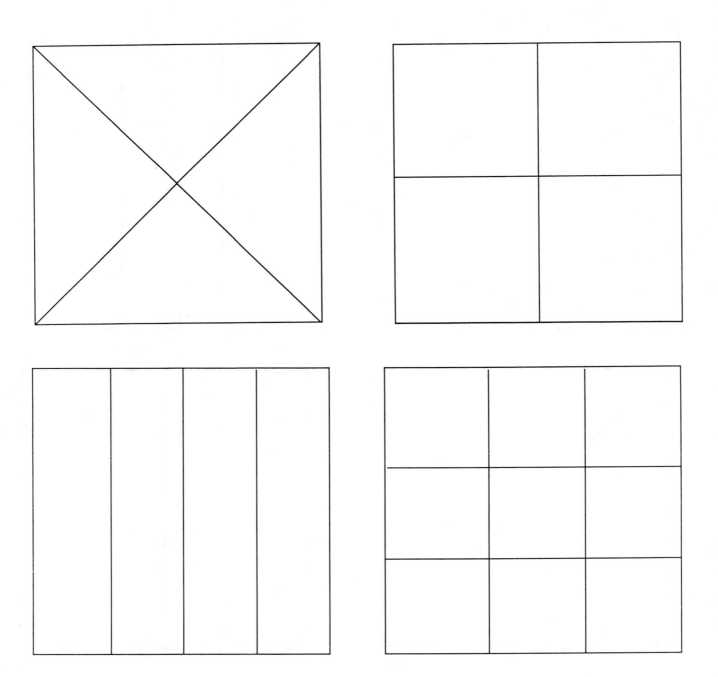

Fractional Parts:
Use With Layered Look Book

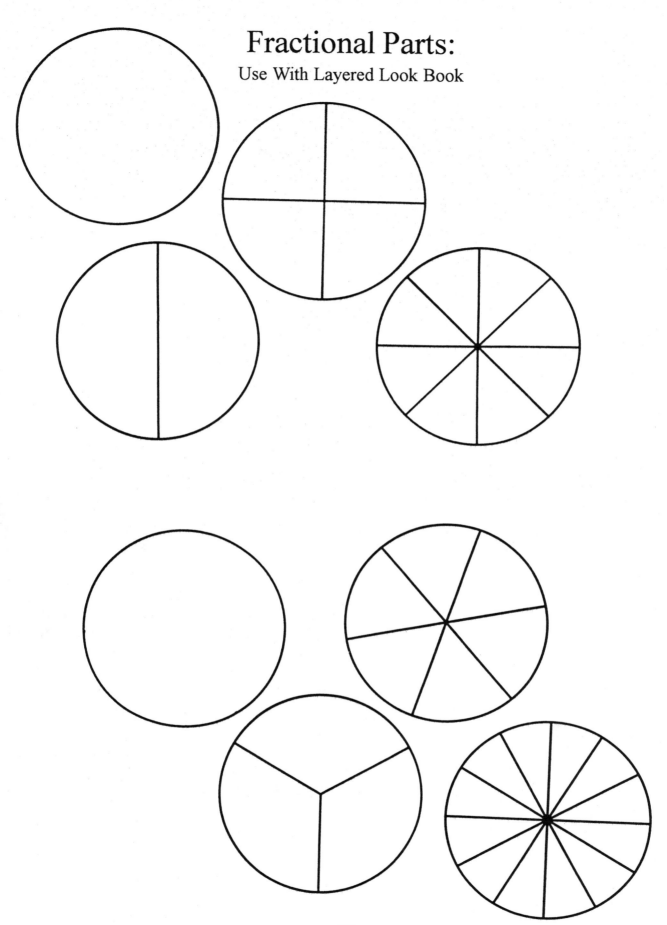

Classroom Fraction Pie: Whole

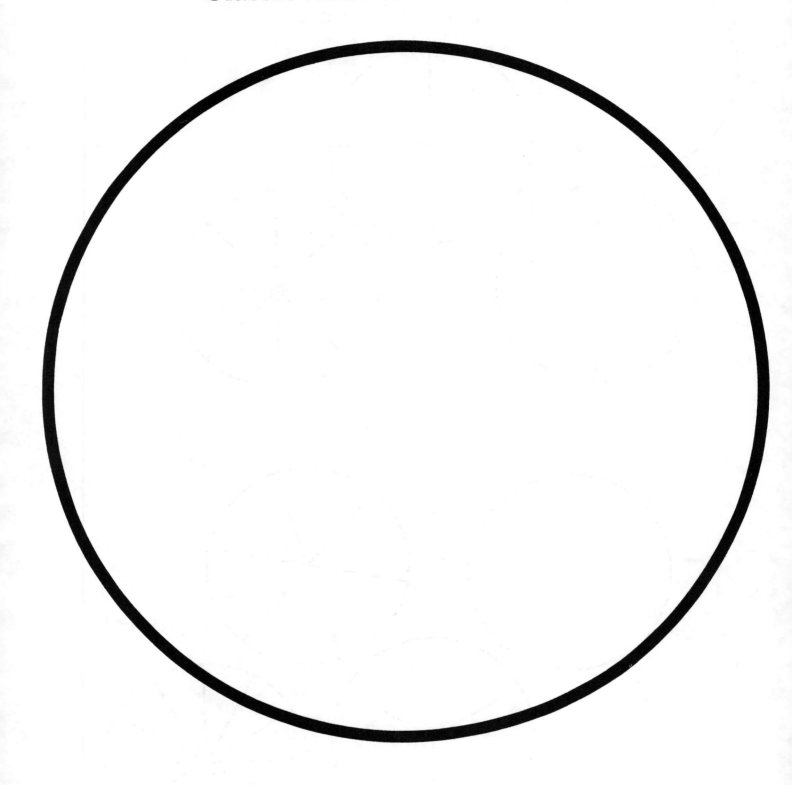

Classroom Fraction Pie: Halves

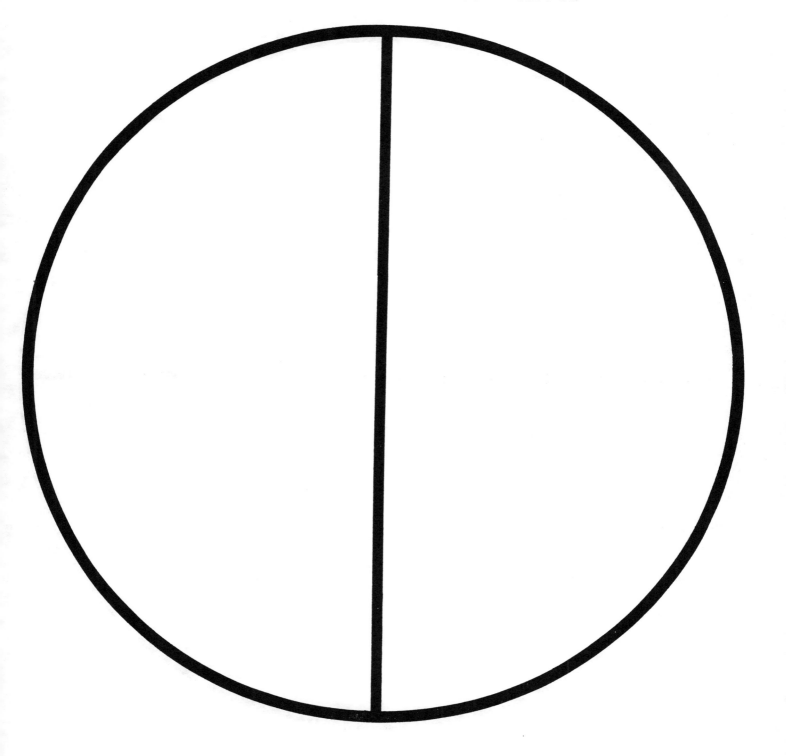

Classroom Fraction Pie: Quarters

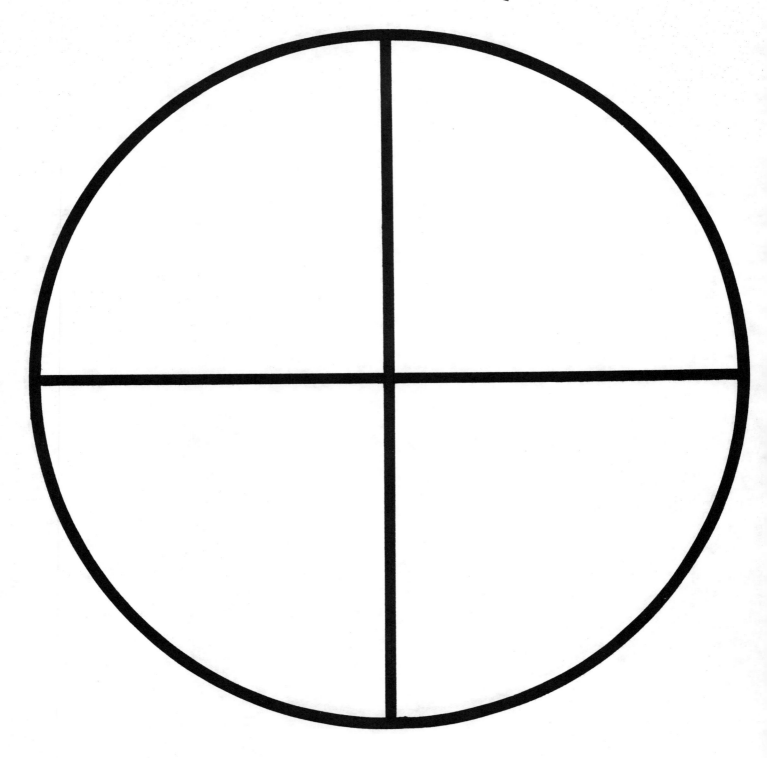

Classroom Fraction Pie: Thirds

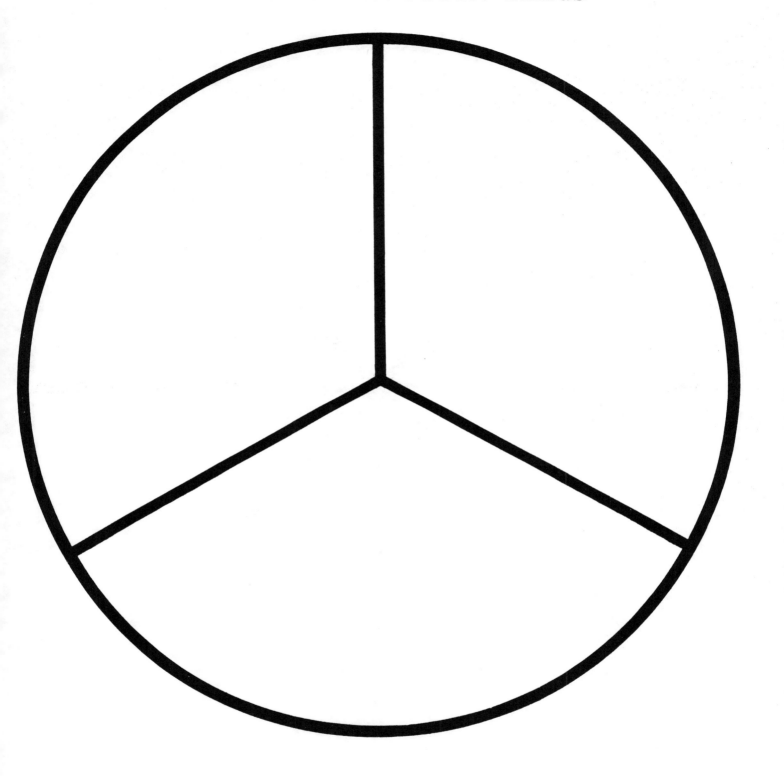

Classroom Fraction Pie: Eighths

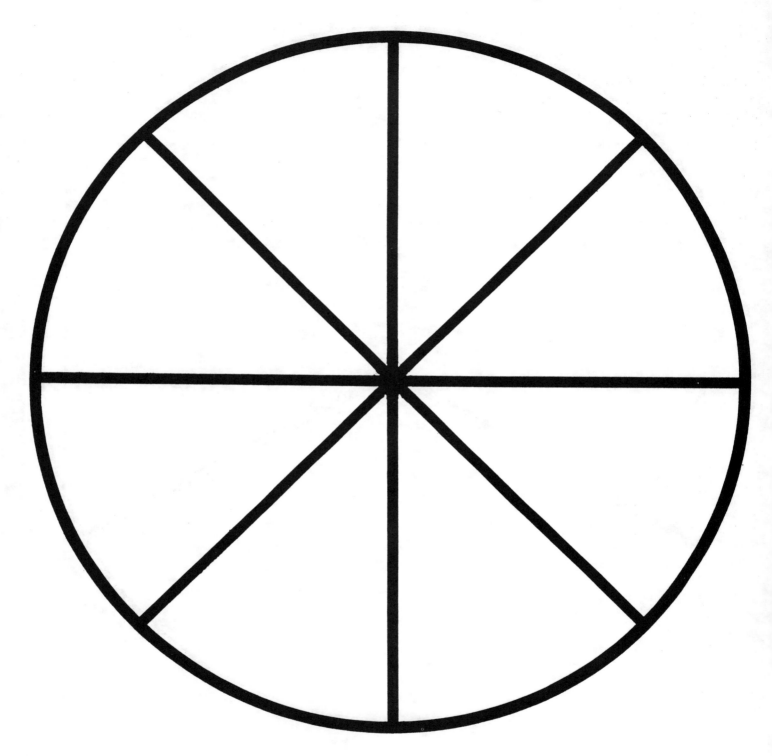

Fractional Parts: Musical Notes

Use With Layered Look Book

Whole Note

1/2 Note

1/2 Note

1/4 Note

1/4 note

1/4 Note

1/4 Note

1/8 Note

1/8 Note

1/8 Note

1/8 Note

1/8Note

1/8 Note

1/8 Note

1/8 Note

Fractional Parts: Moon

Use With Layered Look Book

FULL MOON

1/2 MOON

1/2 MOON

Football Fractional Parts

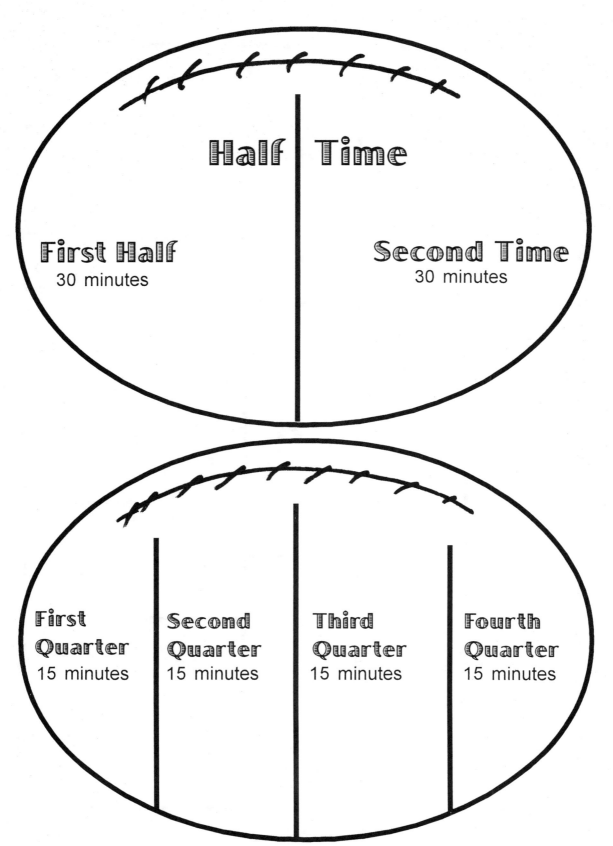

Half Time

First Half
30 minutes

Second Time
30 minutes

First Quarter
15 minutes

Second Quarter
15 minutes

Third Quarter
15 minutes

Fourth Quarter
15 minutes

Mix and Match Fraction Bars

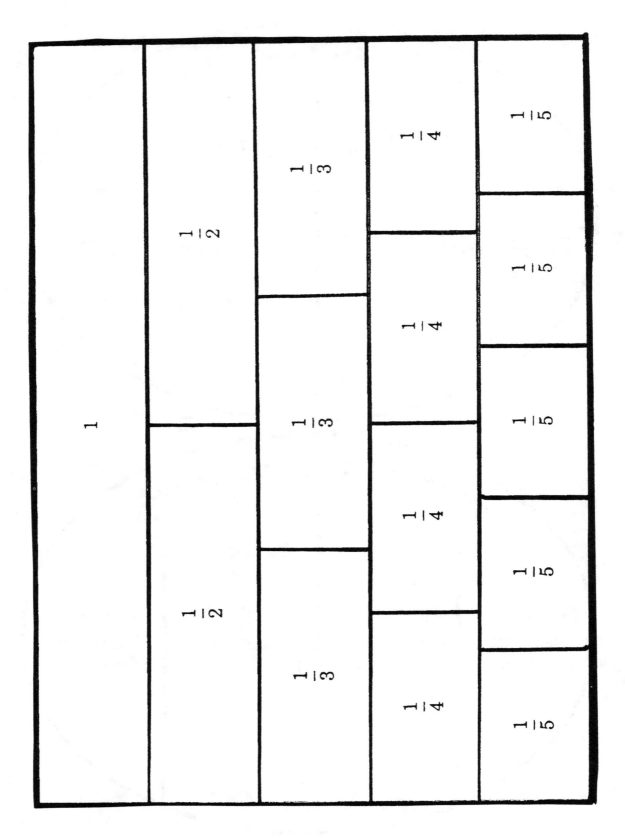

Mix and Match Fraction Bars

$\frac{1}{6}$	$\frac{1}{8}$	$\frac{1}{9}$	$\frac{1}{10}$	$\frac{1}{12}$
			$\frac{1}{10}$	$\frac{1}{12}$
$\frac{1}{6}$	$\frac{1}{8}$	$\frac{1}{9}$		$\frac{1}{12}$
	$\frac{1}{8}$	$\frac{1}{9}$	$\frac{1}{10}$	$\frac{1}{12}$
$\frac{1}{6}$		$\frac{1}{9}$		$\frac{1}{12}$
	$\frac{1}{8}$		$\frac{1}{10}$	$\frac{1}{12}$
$\frac{1}{6}$	$\frac{1}{8}$	$\frac{1}{9}$	$\frac{1}{10}$	$\frac{1}{12}$
		$\frac{1}{9}$	$\frac{1}{10}$	$\frac{1}{12}$
$\frac{1}{6}$	$\frac{1}{8}$			$\frac{1}{12}$
		$\frac{1}{9}$	$\frac{1}{10}$	$\frac{1}{12}$
$\frac{1}{6}$	$\frac{1}{8}$	$\frac{1}{9}$	$\frac{1}{10}$	$\frac{1}{12}$

Geometry

Concentric Circles

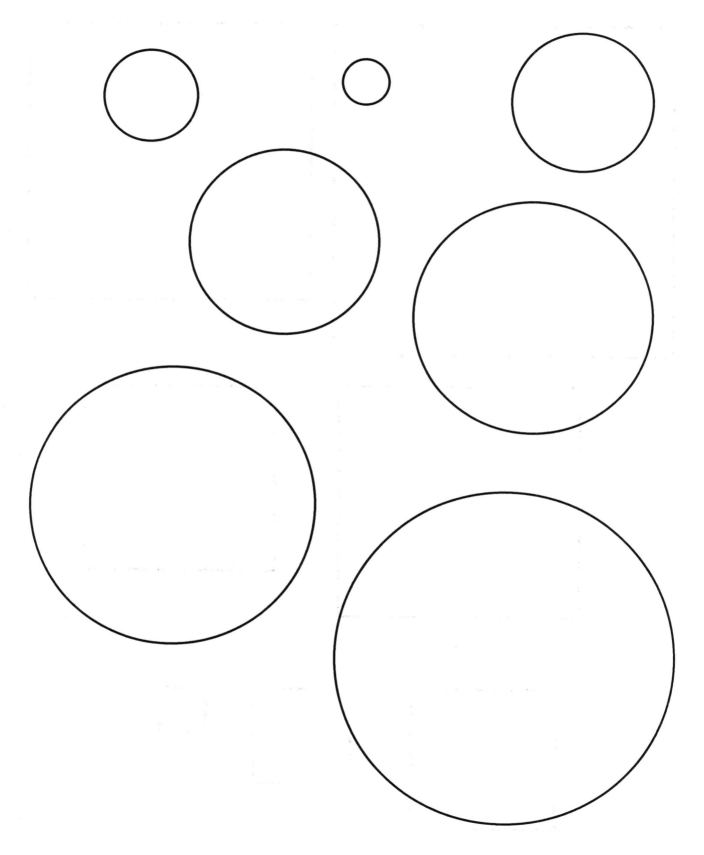

Concentric Squares

Concentric Shapes

Two-Dimensional Shapes

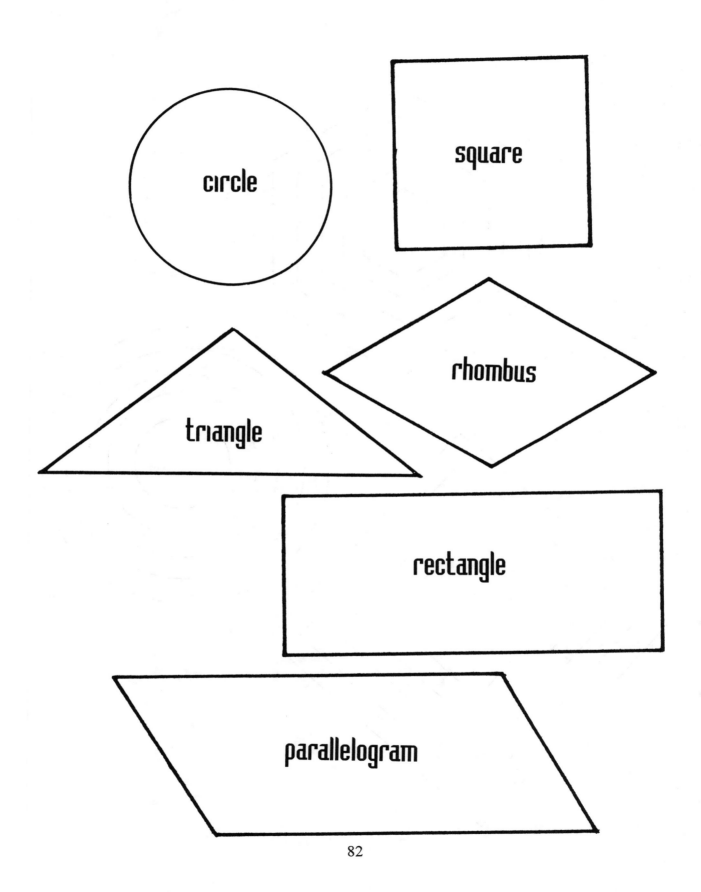

circle

square

triangle

rhombus

rectangle

parallelogram

Three-Dimensional Shapes

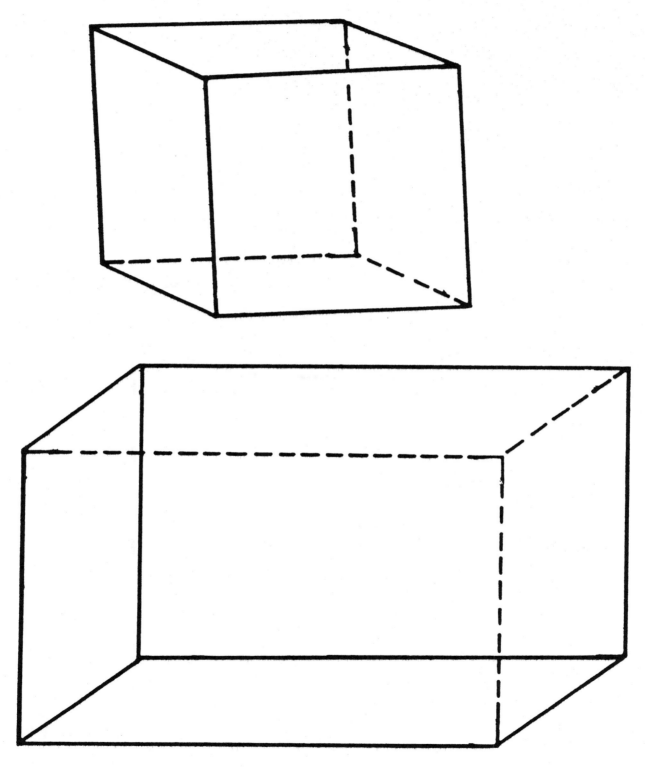

Three-Dimensional Shapes

Three-Dimensional Shapes

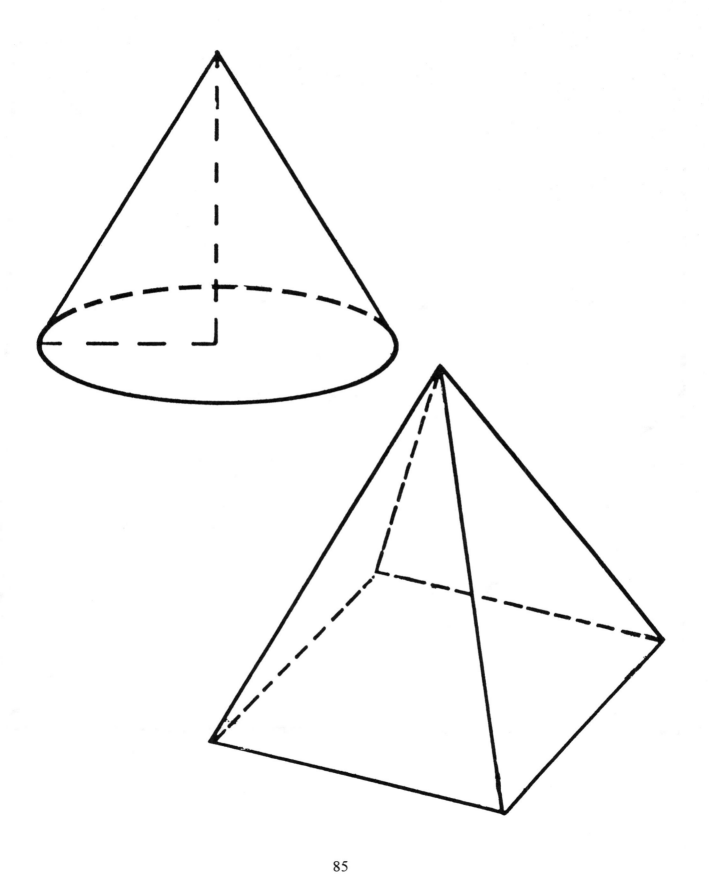

Tangram

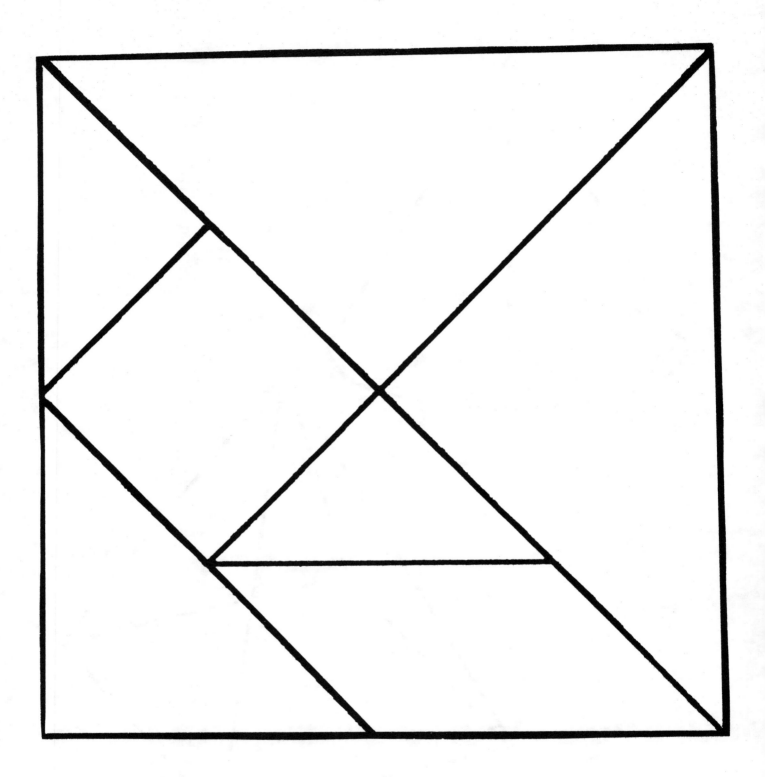

3-D Geometrical Shape Model
Cube

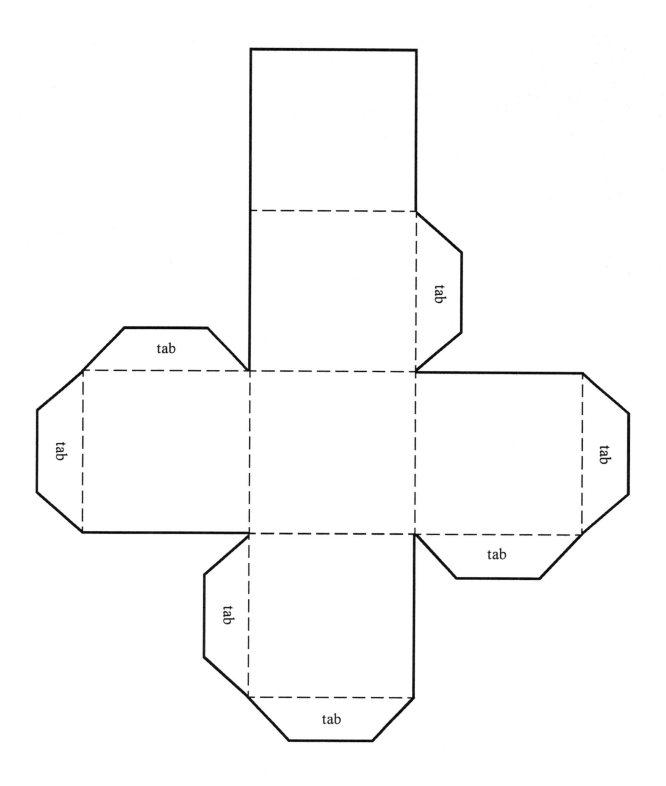

3-D Geometrical Shape Model
Box

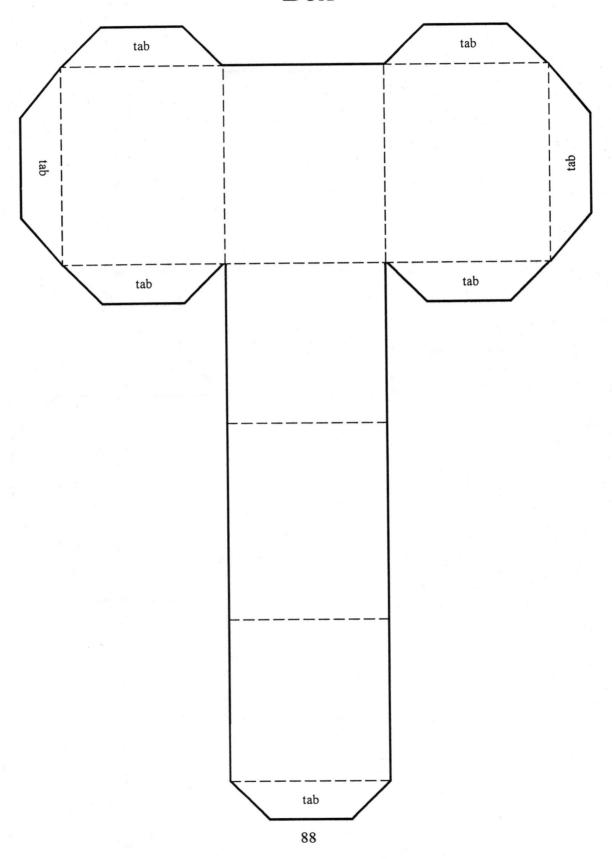

tab

tab

tab

tab

tab

tab

tab

3-D Geometrical Shape Model
Tetrahedron

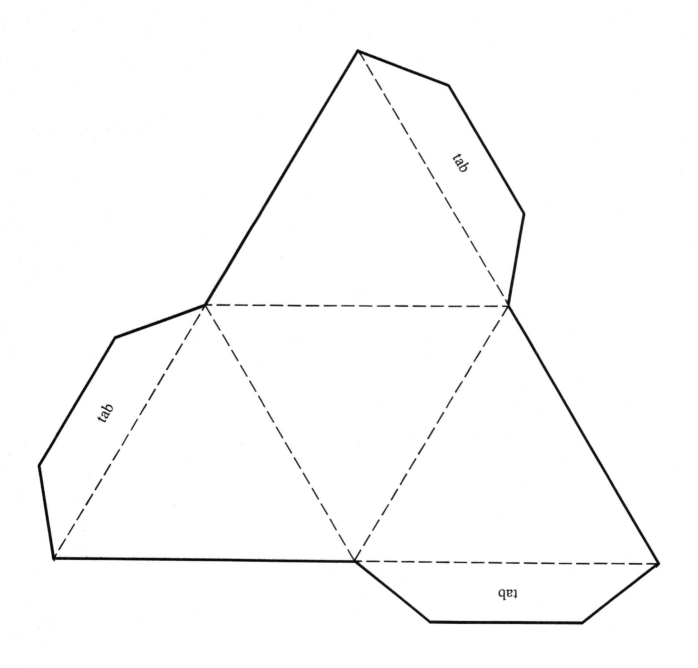

Protractors

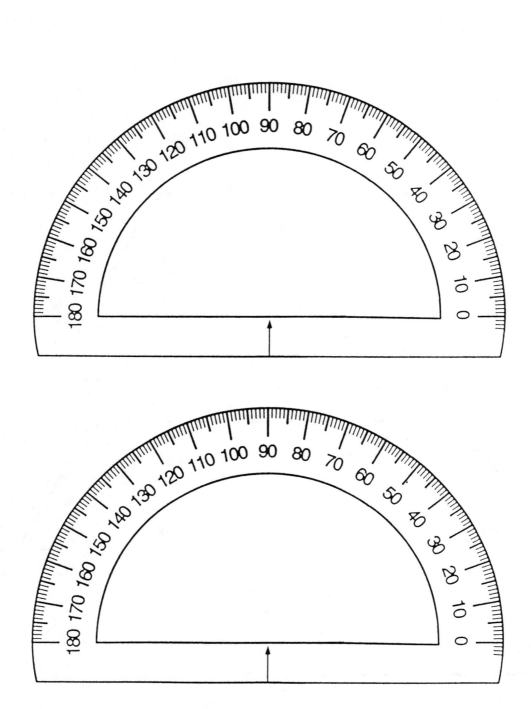

Angles

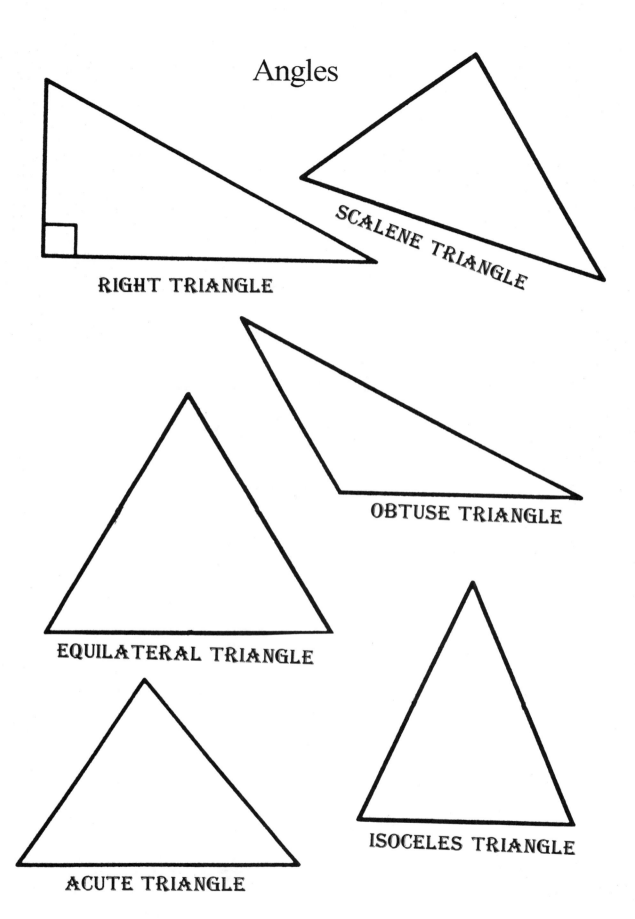

RIGHT TRIANGLE

SCALENE TRIANGLE

OBTUSE TRIANGLE

EQUILATERAL TRIANGLE

ACUTE TRIANGLE

ISOCELES TRIANGLE

Numeration,
Place Value
and
Counting

Hindu-Arabic Counting System

Units

ones	1
tens	10
hundreds	100

Thousands

ones	1,000
tens	10,000
hundreds	100,000

Millions

ones	1,000,000
tens	10,000,000
hundreds	100,000,000

Billions

ones	1,000,000,000
tens	10,000,000,000
hundreds	100,000,000,000

Trillions

ones	1,000,000,000,000
tens	10,000,000,000,000
hundreds	100,000,000,000,000

Quadrillions

ones	1,000,000,000,000,000
tens	10,000,000,000,000,000
hundreds	100,000,000,000,000,000

Quintillions   1,000,000,000,000,000,000

Hundred's

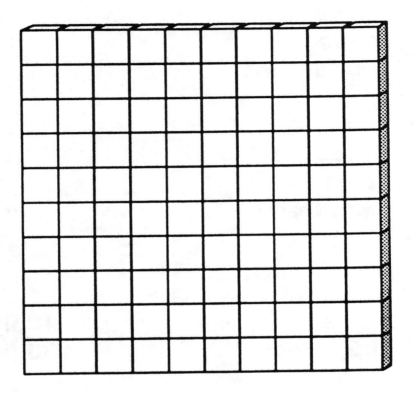

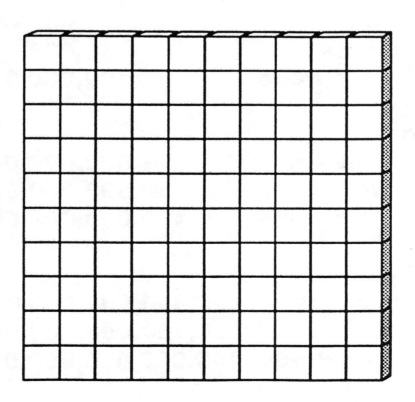

Tens and Ones

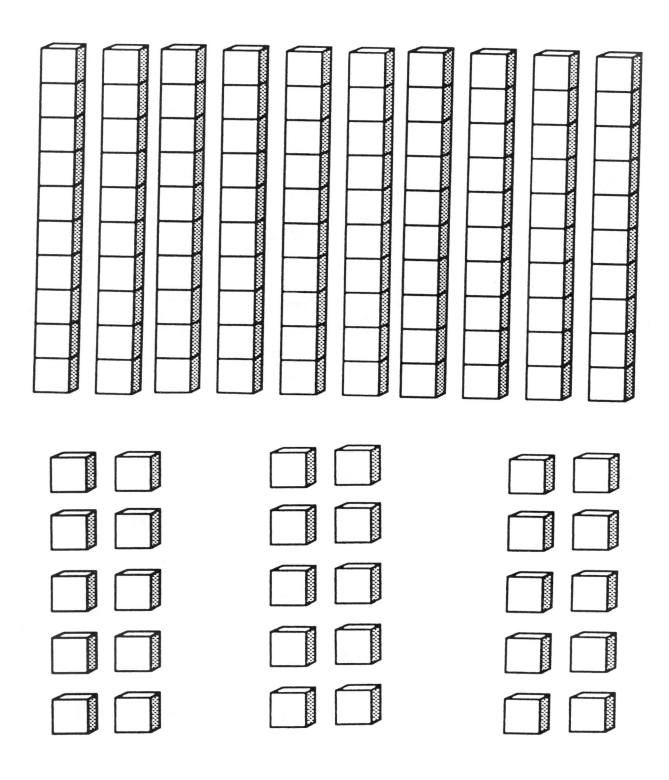

Small Place Value Graphics

Student Place Value Graphics

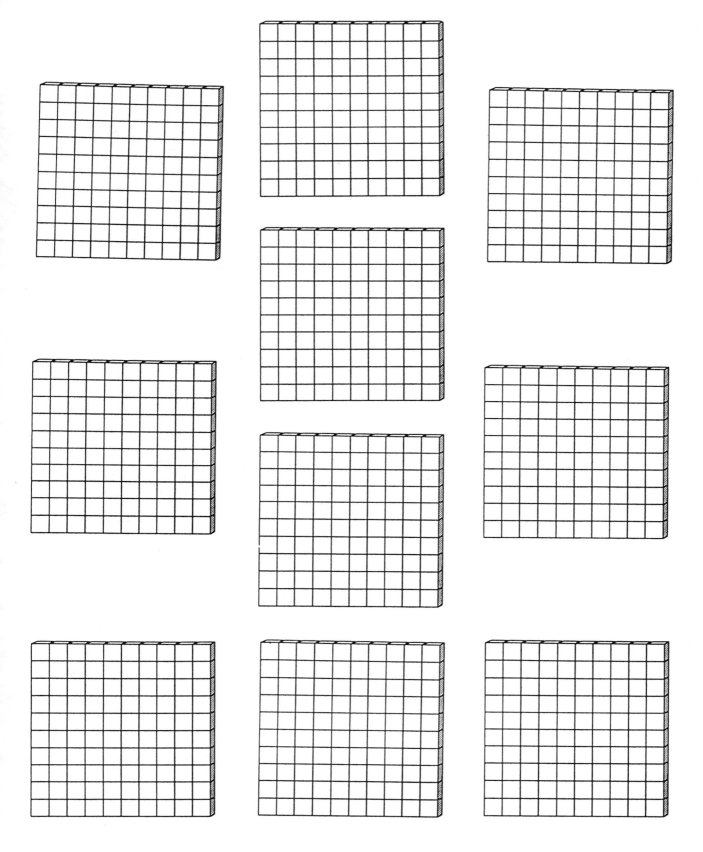

One Hundred

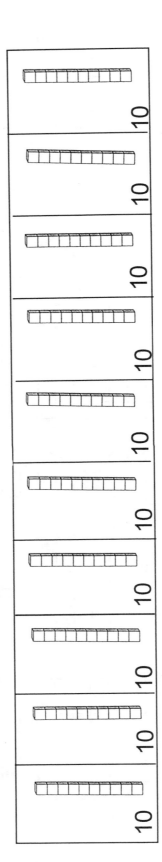

10 10 10 10 10 10 10 10 10 10

96	97	98	99	100
91	92	93	94	95
86	87	88	89	90
81	82	83	84	85
76	77	78	79	80
71	72	73	74	75
66	67	68	69	70
61	62	63	64	65
56	57	58	59	60
51	52	53	54	55
46	47	48	49	50
41	42	43	44	45
36	37	38	39	40
31	32	33	34	35
26	27	28	29	30
21	22	23	24	25
16	17	18	19	20
11	12	13	14	15
6	7	8	9	10
1	2	3	4	5

Whole Number Line

hundred thousands	ten thousands	thousands	hundreds	tens	ones
					●

hundred thousands	ten thousands	thousands	hundreds	tens	ones
					●

hundred thousands	ten thousands	thousands	hundreds	tens	ones
					●

Decimal Line

ones	tenths	hundredths	thousandths	ten thousandths

ones	tenths	hundredths	thousandths	ten thousandths

ones	tenths	hundredths	thousandths	ten thousandths

1	one	first
2	two	second
3	three	third
4	four	fourth
5	five	fifth
6	six	sixth
7	seven	seventh
8	eight	eighth
9	nine	ninth
10	ten	tenth

11	eleven	eleventh
12	twelve	twelvth
13	thirteen	thirteenth
14	fourteen	fourteenth
15	fifteen	fifteenth
16	sixteen	sixteenth
17	seventeen	seventeenth
18	eighteen	eighteenth
19	nineteen	ninteenth
20	twenty	twentieth

1	1st	11	11th
2	2nd	12	12th
3	3rd	13	13th
4	4th	14	14th
5	5th	15	15th
6	6th	16	16th
7	7th	17	17th
8	8th	18	18th
9	9th	19	19th
10	10th	20	20th

2	3	5	7
11	13	17	19
23	29	31	37
41	43	47	53
59	61	67	71
73	79	83	89
97	101	103	107
109	113	127	131
137	139	149	151
157	163	167	173

Composite Numbers

4	6	8	9
10	12	14	15
16	18	20	21
22	24	25	26
27	28	30	32
33	34	35	36
38	39	40	42
44	45	46	48
49	50	51	52
54	55	56	57

58	60	62	63
64	65	66	68
69	70	72	74
75	76	77	78
80	81	82	84
85	86	87	88
90	91	92	93
94	95	96	98
99	100	102	104
105	106	108	110

I	1
II	2
III	3
IV	4
V	5
VI	6
VII	7
VIII	8
IX	9
X	10

XI	11
XII	12
XIII	13
XIV	14
XV	15
XVI	16
XVII	17
XVIII	18
XIX	19
XX	20

XXX	30
XL	40
L	50
LX	60
LXX	70
LXXX	80
XC	90
C	100
D	500
M	1000

Skip Counting Cards:

Even Numbers-2's
Odd Numbers-3's
and
4's-12's

2	22
4	24
6	26
8	28
10	30
12	32
14	34
16	36
18	38
20	40

3	33
6	36
9	39
12	42
15	45
18	48
21	51
24	54
27	57
30	60

4	44
8	48
12	52
16	56
20	60
24	64
28	68
32	72
36	76
40	80

5	55
10	60
15	65
20	70
25	76
30	80
35	85
40	90
45	95
50	100

6	66
12	72
18	78
24	84
30	90
36	96
42	102
48	108
54	114
60	120

7	77
14	84
21	91
28	98
35	105
42	112
49	119
56	126
63	133
70	140

8	88
16	96
24	104
32	112
40	120
48	128
56	136
64	144
72	152
80	160

9	99
18	108
27	117
36	126
45	135
54	144
63	153
72	162
81	171
90	180

10	110
20	120
30	130
40	140
50	150
60	160
70	170
80	180
90	190
100	200

11	121
22	132
33	143
44	154
55	165
66	176
77	187
88	198
99	209
110	220

12	132
24	144
36	156
48	168
60	180
72	192
84	204
96	216
108	228
120	240

Large Number Array

1	2	3	4	5	6	7	8	9	10
11	12	13	14	15	16	17	18	19	20
21	22	23	24	25	26	27	28	29	30
31	32	33	34	35	36	37	38	39	40
41	42	43	44	45	46	47	48	49	50
51	52	53	54	55	56	57	58	59	60
61	62	63	64	65	66	67	68	69	70
71	72	73	74	75	76	77	78	79	80
81	82	83	84	85	86	87	88	89	90
91	92	93	94	95	96	97	98	99	100

Small Number Array

1	2	3	4	5	6	7	8	9	10
11	12	13	14	15	16	17	18	19	20
21	22	23	24	25	26	27	28	29	30
31	32	33	34	35	36	37	38	39	40
41	42	43	44	45	46	47	48	49	50
51	52	53	54	55	56	57	58	59	60
61	62	63	64	65	66	67	68	69	70
71	72	73	74	75	76	77	78	79	80
81	82	83	84	85	86	87	88	89	90
91	92	93	94	95	96	97	98	99	100

Assignment:

Basic Operations:

Addition,
Subtraction,
Multiplication,
and Division Facts

Student Addition Table

+	0	1	2	3	4	5	6	7	8	9	10	11	12
0	0	1	2	3	4	5	6	7	8	9	10	11	12
1	1	2	3	4	5	6	7	8	9	10	11	12	13
2	2	3	4	5	6	7	8	9	10	11	12	13	14
3	3	4	5	6	7	8	9	10	11	12	13	14	15
4	4	5	6	7	8	9	10	11	12	13	14	15	16
5	5	6	7	8	9	10	11	12	13	14	15	16	17
6	6	7	8	9	10	11	12	13	14	15	16	17	18
7	7	8	9	10	11	12	13	14	15	16	17	18	19
8	8	9	10	11	12	13	14	15	16	17	18	19	20
9	9	10	11	12	13	14	15	16	17	18	19	20	21
10	10	11	12	13	14	15	16	17	18	19	20	21	22
11	11	12	13	14	15	16	17	18	19	20	21	22	23
12	12	13	14	15	16	17	18	19	20	21	22	23	24

Symbols

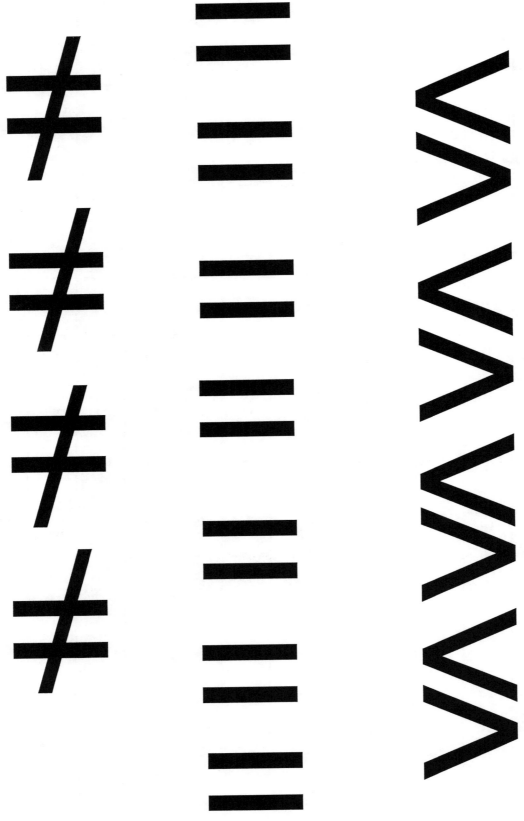

Student Addition Flashcards

1 +0	1 +1	1 +2	1 +3
1 +4	1 +5	1 +6	1 +7
1 +8	1 +9	1 +10	1 +11
2 +0	2 +1	2 +2	2 +3

Student Addition Flashcards

2 +4	2 +5	2 +6	2 +7
2 +8	2 +9	2 +10	2 +11
3 +0	3 +1	3 +2	3 +3
3 +4	3 +5	3 +6	3 +7

Student Addition Flashcards

3 +8	3 +9	3 +10	3 +11
4 +0	4 +1	4 +2	4 +3
4 +4	4 +5	4 +6	4 +7
4 +8	4 +9	4 +10	4 +11

Student Addition Flashcards

5 +0	5 +1	5 +2	5 +3
5 +4	5 +5	5 +6	5 +7
5 +8	5 +9	5 +10	5 +11
6 +0	6 +1	6 +2	6 +3

Student Addition Flashcards

6 +4	6 +5	6 +6	6 +7
6 +8	6 +9	6 +10	6 +11
7 +0	7 +1	7 +2	7 +3
7 +4	7 +5	7 +6	7 +7

Student Addition Flashcards

7 +8	7 +9	7 +10	7 +11
8 +0	8 +1	8 +2	8 +3
8 +4	8 +5	8 +6	8 +7
8 +8	8 +9	8 +10	8 +11

Student Subtraction Flashcards

9 $+0$	9 $+1$	9 $+2$	9 $+3$
9 $+4$	9 $+5$	9 $+6$	9 $+7$
9 $+8$	9 $+9$	9 $+10$	9 $+11$

Student Subtraction Flashcards

0 − 0	1 − 0	1 − 1	2 − 0
2 − 1	2 − 2	3 − 0	3 − 1
3 − 2	3 − 3	4 − 0	4 − 1
4 − 2	4 − 3	4 − 4	5 − 0

Student Subtraction Flashcards

5 -1	5 -2	5 -3	5 -4
5 -5	6 -0	6 -1	6 -2
6 -3	6 -4	6 -5	6 -6
7 -0	7 -1	7 -2	7 -3

Student Subtraction Flashcards

7 −4	7 −5	7 −6	7 −7
8 −0	8 −1	8 −2	8 −3
8 −4	8 −5	8 −6	8 −7
8 −8	9 −0	9 −1	9 −2

Student Subtraction Flashcards

9 -3	9 -4	9 -5	9 -6
9 -7	9 -8	9 -9	10 -0
10 -1	10 -2	10 -3	10 -4
10 -5	10 -6	10 -7	10 -8

Student Multiplication Table

×	0	1	2	3	4	5	6	7	8	9	10	11	12
0	0	0	0	0	0	0	0	0	0	0	0	0	0
1	0	1	2	3	4	5	6	7	8	9	10	11	12
2	0	2	4	6	8	10	12	14	16	18	20	22	24
3	0	3	6	9	12	15	18	21	24	27	30	33	36
4	0	4	8	12	16	20	24	28	32	36	40	44	48
5	0	5	10	15	20	25	30	35	40	45	50	55	60
6	0	6	12	18	24	30	36	42	48	54	60	66	72
7	0	7	14	21	28	35	42	49	56	63	70	77	84
8	0	8	16	24	32	40	48	56	64	72	80	88	96
9	0	9	18	27	36	45	54	63	72	81	90	99	108
10	0	10	20	30	40	50	60	70	80	90	100	110	120
11	0	11	22	33	44	55	66	77	88	99	110	121	132
12	0	12	24	36	48	60	72	84	96	108	120	132	144

Student Multiplication Flashcards

2 x4	2 x5	2 x6	2 x7
2 x8	2 x9	2 x10	2 x11
3 x0	3 x1	3 x2	3 x3
3 x4	3 x5	3 x6	3 x7

Student Multiplication Flashcards

3 x8	3 x9	3 x10	3 x11
4 x0	4 x1	4 x2	4 x3
4 x4	4 x5	4 x6	4 x7
4 x8	4 x9	4 x10	4 x11

Student Multiplication Flashcards

5 x0	5 x1	5 x2	5 x3
5 x4	5 x5	5 x6	5 x7
5 x8	5 x9	5 x10	5 x11
6 x0	6 x1	6 x2	6 x3

Student Multiplication Flashcards

6 x4	6 x5	6 x6	6 x7
6 x8	6 x9	6 x10	6 x11
7 x0	7 x1	7 x2	7 x3
7 x4	7 x5	7 x6	7 x7

Student Multiplication Flashcards

7 x8	7 x9	7 x10	7 x11
8 x0	8 x1	8 x2	8 x3
8 x4	8 x5	8 x6	8 x7
8 x8	8 x9	8 x10	8 x11

Student Multiplication Flashcards

9 x0	9 x1	9 x2	9 x3
9 x4	9 x5	9 x6	9 x7
9 x8	9 x9	9 x10	9 x11

Student Division Flashcards

1 ÷ 1 =	3 ÷ 3 =
	3 ÷ 1 =
	2 ÷ 2 =
	2 ÷ 1 =

Student Division Flashcards

5÷5 =	6÷6 =	8÷2 =	9÷3 =
5÷1 =	6÷3 =	8÷1 =	9÷1 =
4÷2 =	6÷2 =	7÷7 =	8÷8 =
4÷1 =	6÷1 =	7÷1 =	8÷4 =

Student Division Flashcards

10 ÷ 5 =	12 ÷ 6 =	15 ÷ 3 =	16 ÷ 8 =
10 ÷ 2 =	12 ÷ 3 =	14 ÷ 7 =	16 ÷ 2 =
10 ÷ 1 =	12 ÷ 2 =	14 ÷ 2 =	16 ÷ 4 =
9 ÷ 9 =	10 ÷ 10 =	12 ÷ 4 =	15 ÷ 5 =

Student Division Flashcards

$20 \div 2 =$	$21 \div 3 =$	$24 \div 4 =$	$27 \div 3 =$
$18 \div 6 =$	$20 \div 10 =$	$24 \div 3 =$	$25 \div 5 =$
$18 \div 9 =$	$20 \div 5 =$	$24 \div 2 =$	$24 \div 8 =$
$18 \div 2 =$	$20 \div 4 =$	$21 \div 7 =$	$24 \div 6 =$

Student Division Flashcards

30÷5 =	35÷5 =	42÷7 =	48÷8 =
28÷7 =	35÷7 =	40÷8 =	48÷6 =
28÷4 =	32÷8 =	40÷5 =	49÷7 =
27÷9 =	32÷4 =	36÷6 =	42÷6 =

Student Division Flashcards

$56 \div 7 =$	$64 \div 8 =$		
$54 \div 6 =$	$63 \div 7 =$		
$54 \div 9 =$	$63 \div 9 =$		
$50 \div 5 =$	$56 \div 8 =$	$72 \div 9 =$	

X	1	2	3	4	5	6	7	8	9	10	11	12
1												
2												
3												
4												
5												
6												
7												
8												
9												
10												
11												
12												

Graphs and Grids

Plotting Points

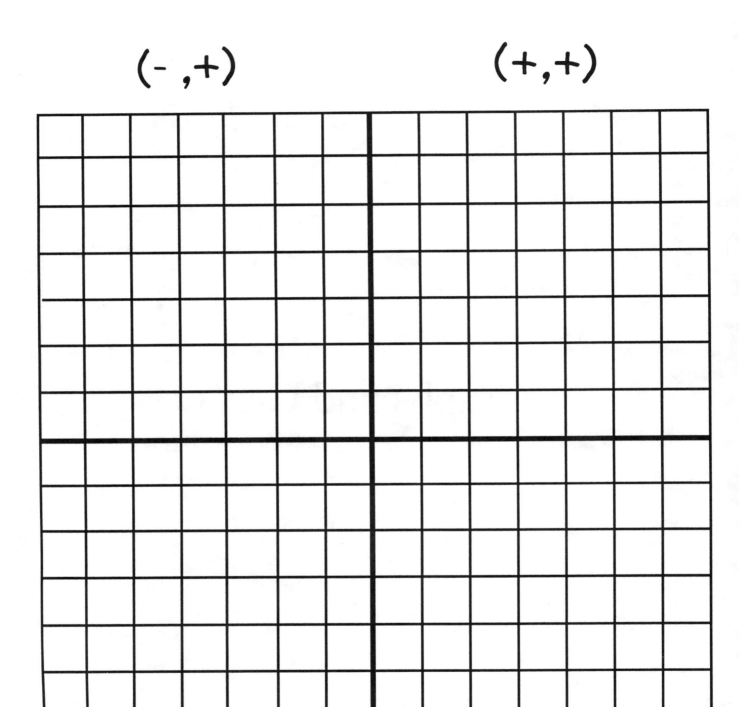

Large Bar Graphs

Large Line Graph

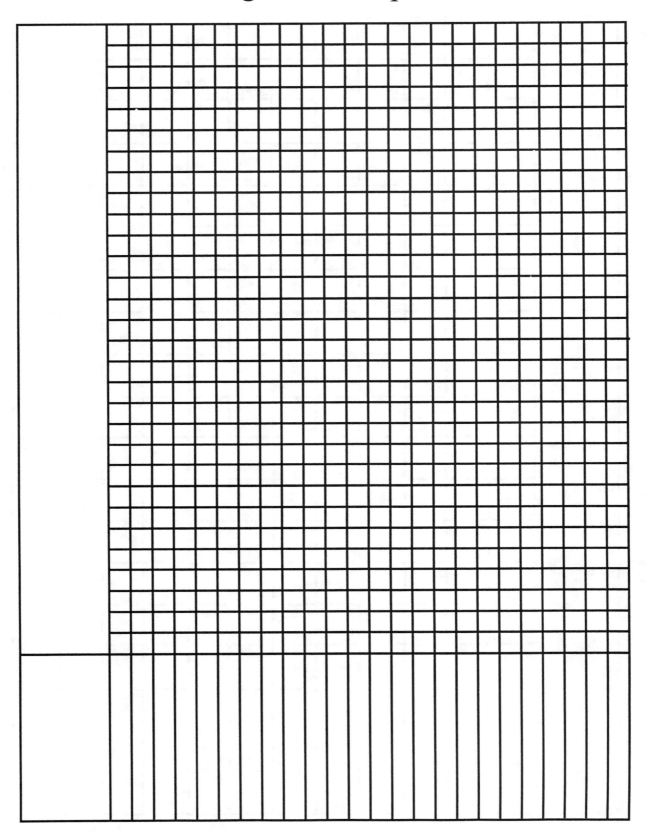

Large Circle Graphs: Percentages

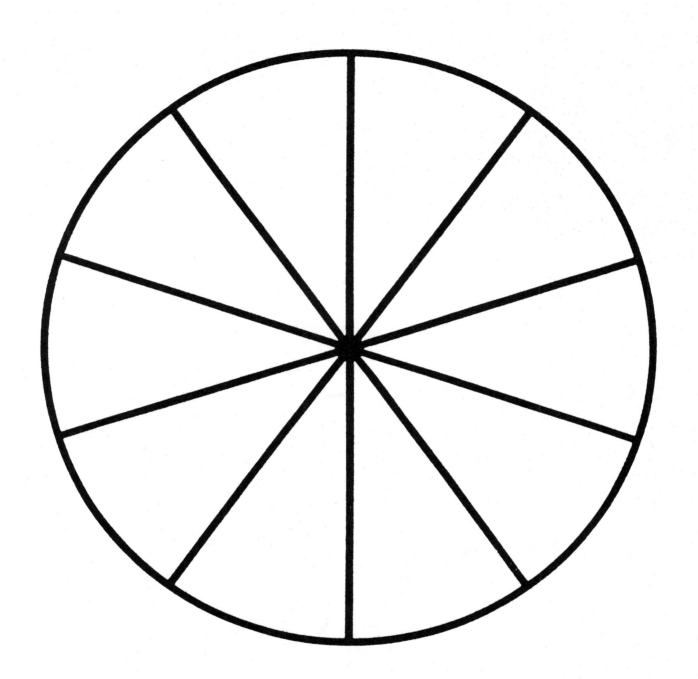

Tenths

Student Circle Graphs: Tenths

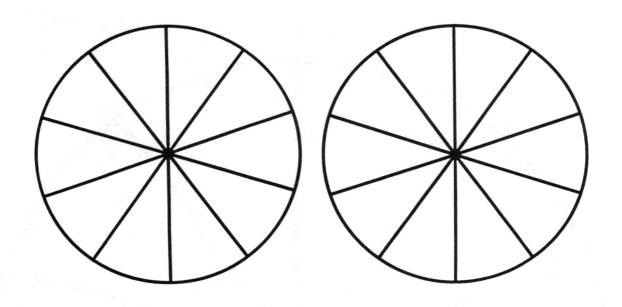

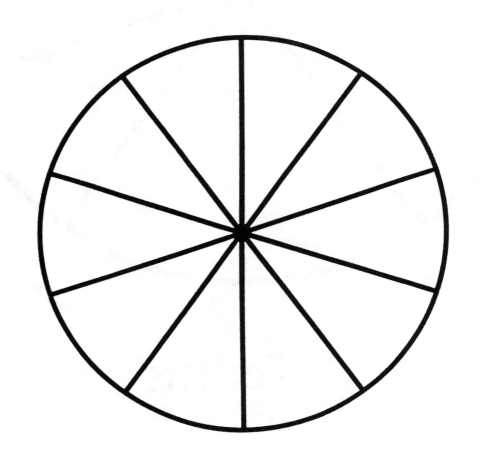

Venn-Diagrams

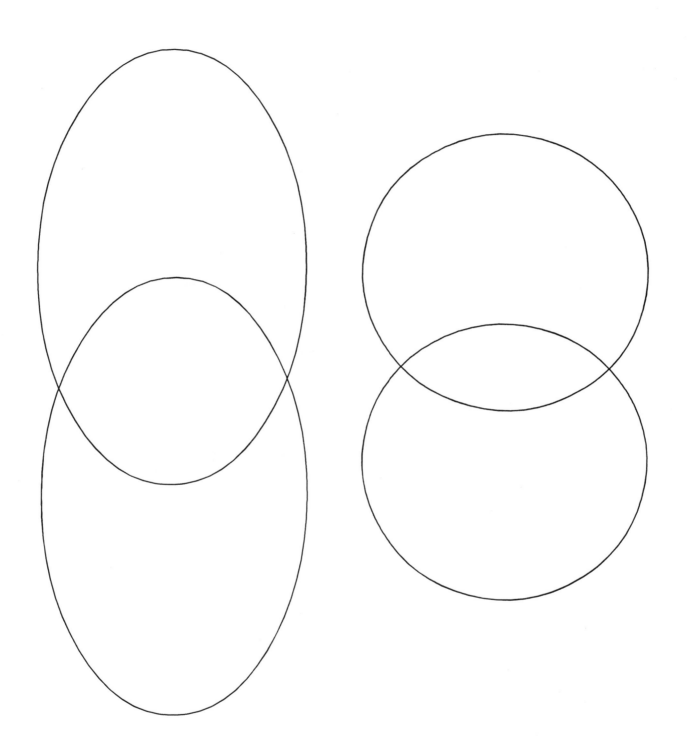

Venn Diagrams

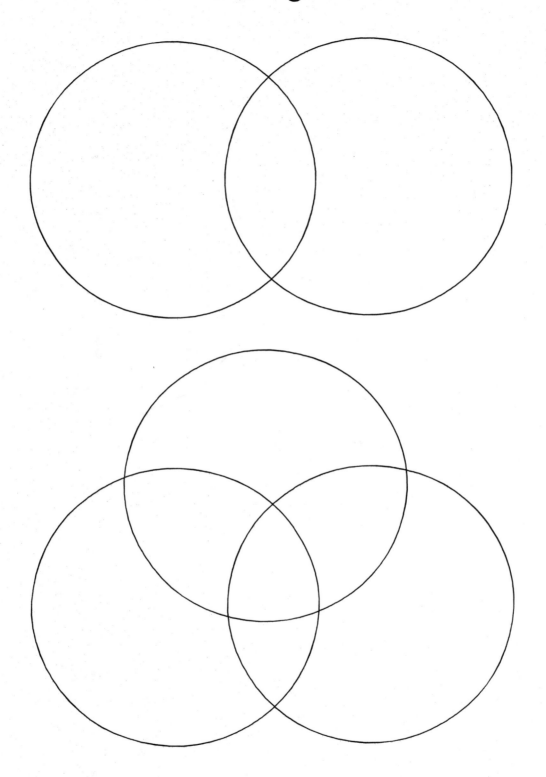

Notes

Dinah Zike's
Dinah-Might Activities
Catalog

Dinah Zike's Big Book Of
PROJECTS
How to design, develop, & make projects, from kindergarten through college

TEACHERS

STUDENTS

HOME SCHOOLING

CHURCHES

CLUBS AND ORGANIZATIONS

By Dinah Zike, M.Ed.

If you can afford only one of Dinah Zike's books, this is it! The award-winning *Big Book of Books and Activities* is used by thousands of teachers and parents internationally. Since its 1989 debut, it has become an education classic, and it is used by experienced teachers, student teachers, and home schooling parents alike. **CCC100 $19.95**

How-To Videos: In each of these one-hour videos, Dinah explains how to use her most popular books. The videos are appropriate for teachers and parents who have attended Dinah's seminars and need a review, or they serve as an introduction to those who have never seen Dinah Zike in person.

This 148-page book is the sequel to Dinah's *Big Book of Books and Activities*. It expands the use of manipulatives presented in *Big Book of Books and Activities*, and it introduces 14 folds not found in that book. The back section contains 64 duplicable pages of Dinah's publishing center graphics. **CCC91 $19.95**

CCC85 Video: *How to Use Dinah Zike's Big Book of Books*
CCC86 Video: *How to Use Dinah Zike's Big Book of Projects*
$12.95 each.

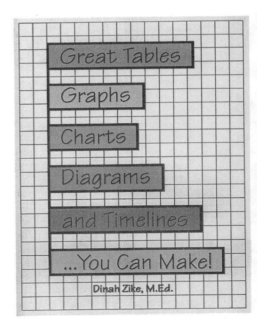

Great Tables

Graphs

Charts

Diagrams

and Timelines

...You Can Make!

Dinah Zike, M.Ed.

For years, teachers have encouraged Dinah to organize her manipulative holiday ideas into a book. *Big Book of Holiday Activities* contains 180 pages of monthly holiday activities and art patterns for the K-5 student. The book also contains lists of story starters, historic dates, and important birthdays to be used as holiday research and writing projects. **CCC83 $24.95**

TO ORDER:
Fax your orders to
(210) 698-0095
Or call toll free
800-99-DINAH
(993-4624)

Standardized tests are filled with tables, graphs, charts, and diagrams. This "Dinah Zike book" presents easy-to-make teaching manipulatives and graphic organizers to teach your students these basic math life skills while preparing them for testing!! This 125-page book is multilevel.
CCC84 $19.95

educational publishing & consulting
p.o. box 690328
san antonio, texas 78269-0328 usa
office: (210) 698-0123
fax: (210) 698-0095
orders only: 1-800-99-dinah
e-mail: cecile@dinah.com
website: http://www.dinah.com

ORDER FORM

Please fax your orders to (210) 698-0095
Or call toll free 800-99-DINAH (993-4624)

(CHECK ONE)

We accept
☐ Visa and
☐ Mastercard

Your Name_____

Address_____

City_____State_____Zip_____

Phone (   )_____Fax (   )_____E-mail_____

Purchase Order Number (If Needed)_____

If Using Purchase Order, Name & Address of School_____

_____

NAME AS IT APPEARS ON CREDIT CARD_____

CREDIT CARD #_____EXP. DATE_____

SIGNATURE_____

Item #	Qty.	Description	Each	Total
CCC 82		BIG BOOK OF MATH GRAPHICS	$19.95	
CCC 83		BIG BOOK OF HOLIDAY ACTIVITIES	$24.95	
CCC 84		GREAT TABLES, GRAPHS, CHARTS, DIAGRAMS, ETC. YOU CAN MAKE!	$19.95	
CCC 85		VIDEO: HOW TO USE THE BIG BOOK OF BOOKS	$12.95	
CCC 86		VIDEO: HOW TO USE THE BIG BOOK OF PROJECTS	$12.95	
CCC 87		TIME TWISTERS: THE LOST NAVIGATORS	$12.95	
CCC 88		TIME TWISTERS: THE HIDDEN CAVERNS	$12.95	
CCC 89		TIME TWISTERS: RAIN FOREST RESCUE	$12.95	
CCC 90		TIME TWISTERS: THE SEARCH FOR T. REX	$12.95	
CCC 91		BIG BOOK OF PROJECTS	$19.95	
CCC 92		THE EARTH SCIENCE BOOK	$12.95	
CCC 93		OLD TESTAMENT SUPPLEMENT TO BIG BOOK OF BOOKS $12.95 SALE!	$8.00	
CCC 94		NEW TESTAMENT SUPPLEMENT TO BIG BOOK OF BOOKS $12.95 SALE!	$8.00	
CCC 100		BIG BOOK OF BOOKS AND ACTIVITIES	$19.95	

SHIPPING & HANDLING:

10% of total order or 8% of orders over $200.
$3.00 minimum S & H on all orders.
3-day, 2-day, & overnight UPS available. Call for prices.
Shipping outside of United States: Call for prices.

Subtotal _____
Shipping & Handling _____
Sales Tax (Texas Residents 7.75%) _____

GRAND TOTAL [   ]

Please call for information on how
to book a Dinah Zike workshop
in your area.
(210) 698-0123

To receive a free catalog,
or to order other books
and materials by Dinah Zike,
please call
1-800-99DINAH

Visit our website at www.Dinah.com
E-mail us at dma@dinah.com

© 1999 Dinah Zike,
Dinah-Might Activities, Inc.
P.O. Box 690328
San Antonio, Texas 78269